Texas Crawdads

Sterling K. Johnson and Nathan K. Johnson

Crawdad Club Designs
College Station, Texas 2008

Published by Crawdad Club Designs
Post Office Box 9651
College Station, Texas 77842-9651

www.texascrawdads.com

Copy editor: Erin Sullivan

Cover design: Kerry Polcsik

Cover drawing: Erin Gelé Johnston

Library of Congress Control Number: 2007908666

ISBN: 978-0-9801103-0-2

Printed in the United States of America
by Grunwald Printing Company, Corpus Christi, Texas.

Cover: *Procambarus steigmani* (female)

Preface

Crawdads delight children more than any other age group. Given opportunity and enough freedom to explore aquatic places, kids soon find and catch the exciting critters. Knowing this, we set about producing a resource aimed at helping young people – especially the elementary through middle school age group – to broaden their experience of crawdads and crawdad catching. We try to help children indirectly, by strengthening and supporting the knowledge of their mentors. This book, then, is written primarily to those desiring to help children learn about crawdads, freshwater habitats and life itself. We also write to a more general readership, to those among us who find pleasure in the experience of natural things. Additionally, because crawdad books are not often written, we integrate a synoptic presentation of scientific information to assist serious students and professional biologists.

Crawdad adventure is a great conversation topic for people of all ages. Nearly everyone older than forty years, it seems, has a crawfish story. Though the many distractions of modern culture have lessened engagement in outdoor adventure, many youngsters too tell great stories about crawfish. Hopefully, this book will serve to rekindle interest in crawfishing – so long a staple in the lives of young Americans.

The subject is presented in its reality, without romanticism or promotion. Experiencing the beauty of crawfishes and their peculiar ways is enough to stir and sustain one's interest. Interest is further enhanced by the lure of the many associated things discovered when experiencing natural aquatic places. The expectant adventurer will find crawfishing to be an easy, inexpensive and almost always successful activity.

Sufficient attention is given to catching methods. Crawdads must be caught for proper study. They cannot be merely watched, as is the case for bird or butterfly study. Crawdad catching provides many of the enjoyments found in similar activities such as fishing, crabbing and natural inventory.

Information is included on how to conduct a number of activities with crawfish, beyond catching them. These activities are enjoyable for both the young and old. In giving advice about the how-to, we do not claim to write as professionals, but only from our experiences. Inclusion of certain activities will probably not gain us accolades with some readers. In our own defense, we merely intended to be comprehensive.

Some information is presented to satisfy general curiosity. Aquaculture and cooking, for example, seem to be subjects that pop up in many crawfish discussions. We have tried to treat these subjects, like others throughout the text, with a simple and brief introductory statement of the essentials.

Most animal images are presented to actual scale. That is, what you see is the actual size of the animal. If an animal's image is not to actual scale, the exception is indicated in text or by placement of a percentage (percentage of actual size) near to the image. Crawfish length is given as total body length, a measurement from the tip of the rostrum to the end of the tail. Lengths are given in inches (an inch equals 2.54 centimeters).

Each species is represented with photos or drawings of multiple specimens. Showing several animals will help readers comprehend the variation expressed in a particular crawfish species.

A complete species presentation is made and an effort is also made to emphasize species groupings. Many users of this book may have little interest in learning the names and ways of all the Texas crawfishes. Thus, we organized the information in a way so that one could know these animals in stages: a crawfish vs. some other aquatic species, a crawfish grouping based on common visual features, and a specific kind of crawfish. Regarding the groupings, we have taken liberty with artificial classification. The book portions dealing with method of identification and guide to species should provide the reader with enough information to make presumptive identifications to the species level.

A consolidation of current knowledge about Texas crawfishes is provided. Texas crawfishes have not been treated comprehensively since 1958 when a paper by Dr. George Penn and Dr. Horton Hobbs Jr. entitled "A contribution toward a knowledge of the crayfishes of Texas Decapoda, Astacidae" was published in the Texas Journal of Science. Subsequent Texas studies included several river basin accounts and a few descriptive works. Texas species were also included in broad treatments of crawfishes either as particular groups or as crawfishes in general. Important among these works was a doctoral dissertation by Dr. Rollin Reimer on crawfishes (which we call grays, prairie burrowers and hairclaws herein) and papers by Drs. Hobbs, Fitzpatrick, Albaugh, Penn, Reimer and others. We were thus fortunate and grateful to have a number of contributions to the knowledge of Texas crawfishes as background for this book.

We began to collect Texas crawfishes in earnest about ten years ago. Though we caught many more, we retained about 1,100 specimens from throughout the state. Non-Texans should appreciate, and this is not a boast, that Texas has a larger land area than the five states to its east. There are 254 counties. We took our boots and nets along on vacations and visits to far away relatives, dipping along the way as patient wives sat charitably in a nearby vehicle. Very enjoyable were companion trips meant to search out the various species. Live specimens were brought home in transfer containers and then moved to more permanent housing. Over time, we learned to make decent photos and drawings, correct identifications and live collections that thrived.

A preface provides a place for authors to express appreciation to those who were supportive in one way or another to the completion of the effort. A number of people made our work much easier. Dr. Kathryn Vaughn and Heather Prestridge of the Texas Cooperative Wildlife Museum, College Station, allowed us space and microscopic equipment to examine specimens. Dr. John Reagor of the Texas Veterinary Medical Diagnostic Laboratory, College Station, provided help in making microscopical scans. Companions made collecting trips more enjoyable and productive. The list of companions includes Becky Johnson, Mary Lou Johnson, Ben Johnson, Josh Lindsay and Trey Wakefield. We enjoyed the experience of veteran crawfishers, some old and some very young, whom we frequently encountered by surprise. They fueled our interest about the critters with exciting stories about their adventures. We are grateful to those among the disinterested who were kind enough to listen anyway as we talked on about crawdads. Finally, the essential cost of the project was borne by the authors. We thank our families for that sacrifice.

Table of Contents

Ready to go.

Introduction

Crawdads

Crawdads — known as crayfish by those who study them and crawfish by those who eat, catch and grow them — are small aquatic animals best known by their claws or pincers. Children and most English-speaking North America call them crawdads or crawdaddies. In this book, the terms crawdad and crawfish are used interchangeably.

Behind a pair of claws, crawdads have four pairs of legs that they use to walk around on the ground and the bottom of water habitats. Their gills are an adaptation for aquatic life. A muscular tail propels them rapidly backward through the water by repeatedly flexing forward under the body. The body has a hard but fragile covering called the exoskeleton or shell. Small North American species grow to nearly 1.5 inch and large species grow to nearly 5 inches (claws are not included in measure).

Crawdad species are distributed on all continents except Africa. In North America, they range from Canada to Central America. The southeastern United States has the greatest concentration of species. Crawdads live where water is continuously available above or below ground. They are generally absent in the western mountains and deserts of the United States except for some permanent streams and mountain lakes.

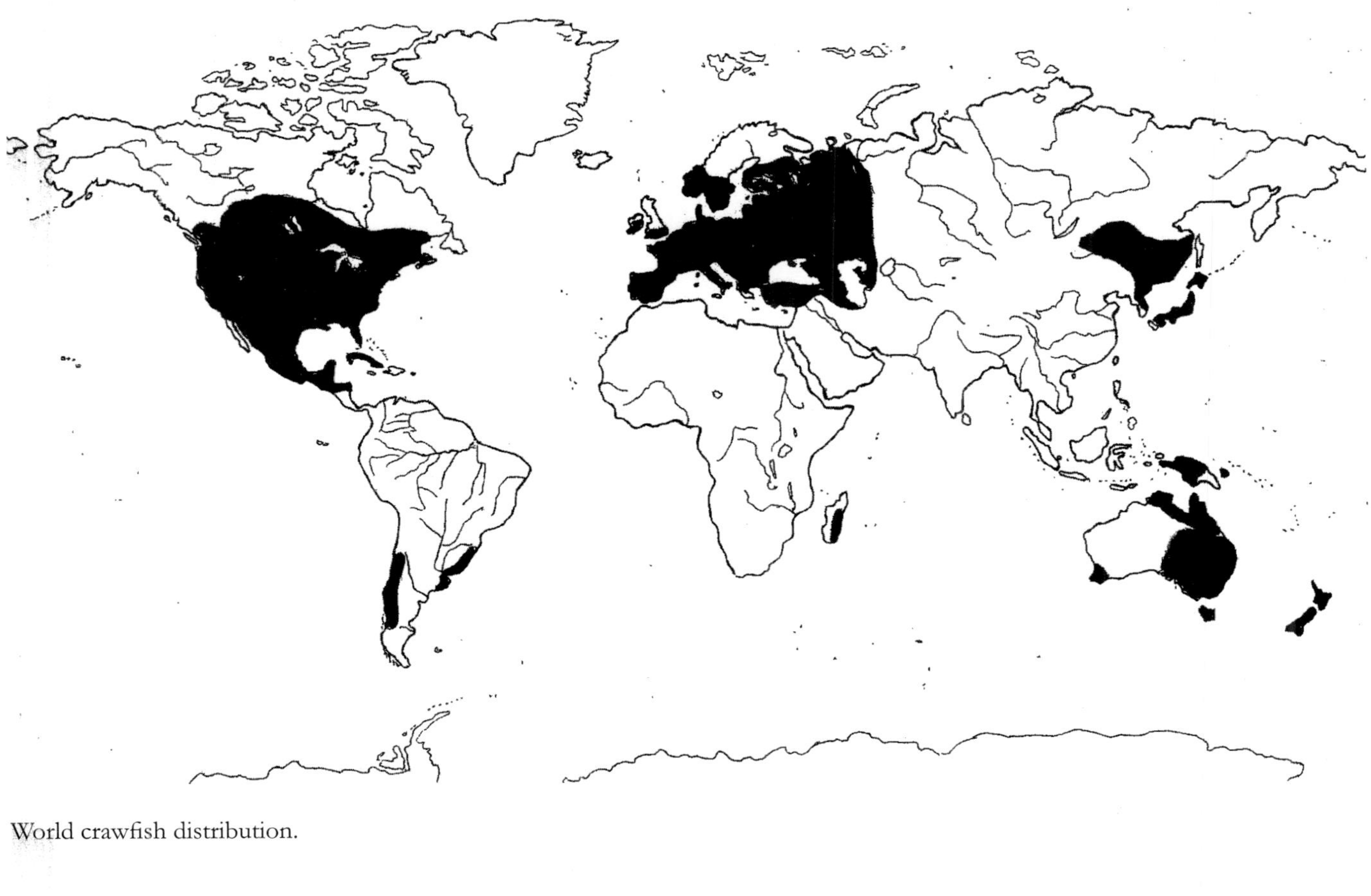

World crawfish distribution.

Crawdads are classified as Crustacea, those animals with a crust-like exoskeleton. As crustaceans, they are classed as decapods, ones with ten legs. The claws make up the first pair of legs and the rest are walking legs. Crawdads differ from other decapods such as shrimps, crabs and lobsters by having a rostrum (a pointed part on top of the head that projects forward from between the eyes) which is not serrated or saw-toothed.

Most decapods live in salt water (marine) environments, but nearly all crawdads and a few shrimps and crabs are freshwater species. Crawdads are a vital part of the food web in aquatic environments and are also important as food for many land animals, people included.

World and North American crawfishes

Of more than 400 freshwater crawfishes that are now known to live throughout the world, most are inhabitants of North America. The North American crawfishes belong to two of the three presently accepted crawfish families: Astacidae and Cambaridae.

The Asticadae is a small family with less than 15 species worldwide. Its North American members (five species) are distributed primarily west of the continental divide in the United States and Canada. There, they live in streams and lakes. Other family members live in Europe, where, to the east, some enter salty water.

North American members of the Cambaridae (more than 270 species) are distributed primarily east of the continental divide. Other crawfishes classified as cambarids (several species) live in Eastern Asia: the Amur River basin of western Russia, Mongolia and northeastern China, the Korean peninsula and Japan.

Distinctively, cambarid males have hooks on the walking legs and well-developed gonopods (modified first appendages of tail). The American females have well-developed sperm receptacles on the underside of their bodies.

Except for human introductions, particularly that of the red crawfish to the American west, geographical separation of the two American families to west and east continues to the present. The red crawfish has also been introduced throughout the world, even to Africa where it now has established permanent residence.

Members (more than 120) of the third family, the Parastacidae, are distributed in the southern hemisphere. There they are found in southern South America, Madagascar, Australia and the larger islands surrounding Australia. Some parastacids are the largest of all crawfish, the oldest individuals weighing as much as eight pounds (3.8 Kg).

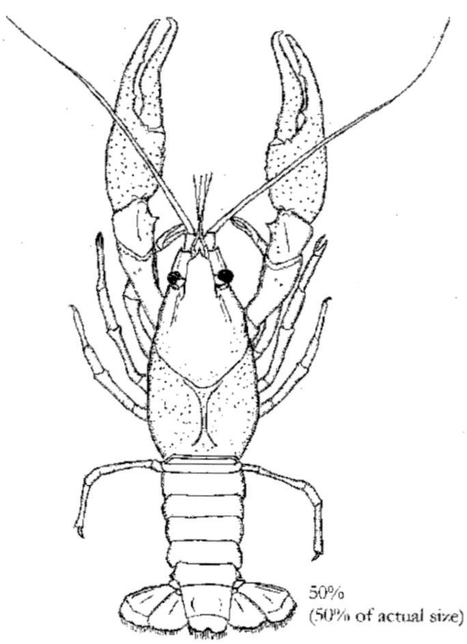

Cambaridae,
Genus *Procambarus*, Texas.

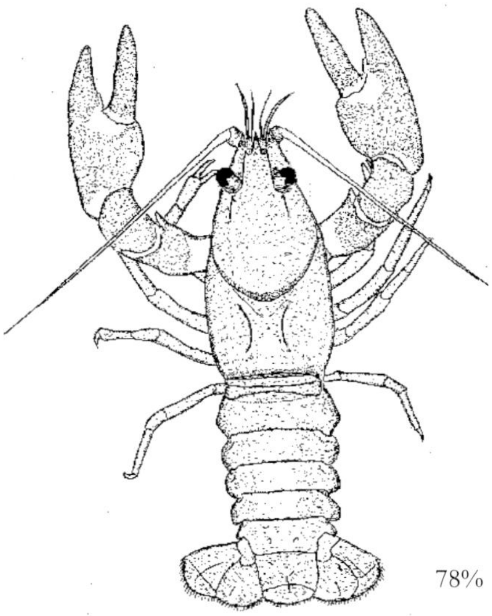

Astacidae,
Genus *Pacifasticus*, California.

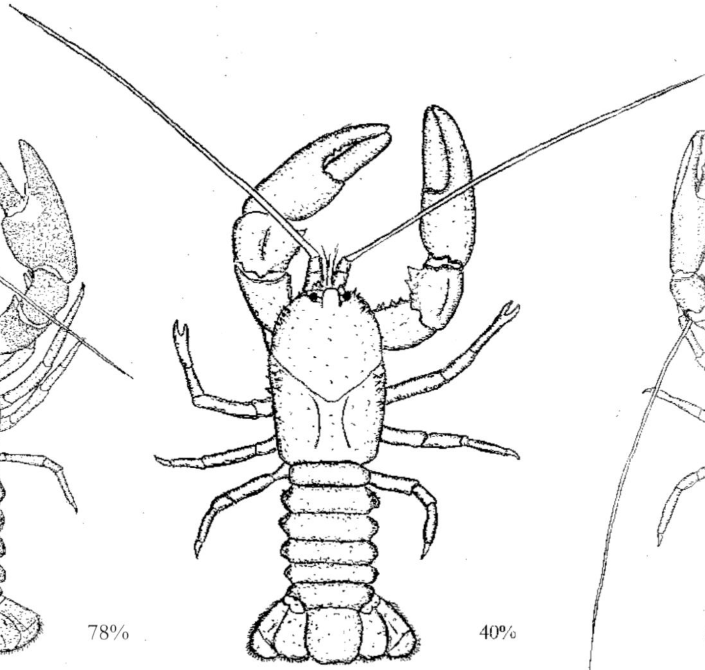

Parastacidae,
Genus *Astacoides*, Madagascar.

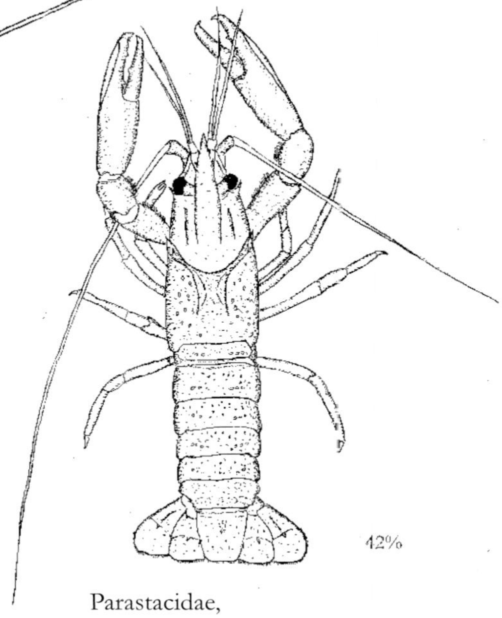

Parastacidae,
Genus *Cherax*, Australia.

All Texas crawfishes belong to the family Cambaridae. Thus, hereafter in this book, all information and illustration concerning crawfishes is limited to that family.

Crawdad Examination

Looking at a crawdad

In order to discuss crawdads in detail and to distinguish one kind from another, it is first necessary to put names to what one sees, that is, the outside structure.

Top view

At first glance from above, a crawfish appears as an elongated body with claws and walking legs extending from beneath. The body is divided into two sections; a segmented, flexible tail (abdomen) and a front portion (cephalothorax) with a non-flexible cover called the carapace. Approximately midway toward the back of the carapace is the cervical groove that extends forward and downward around the sides of the carapace. In front of the groove is the head, and behind it is the thorax. The claws and legs are paired and segmented. The ends of the claws and the first two pairs of walking legs have pinchers. The pinchers have two fingers, the outer one fixed and the other one movable.

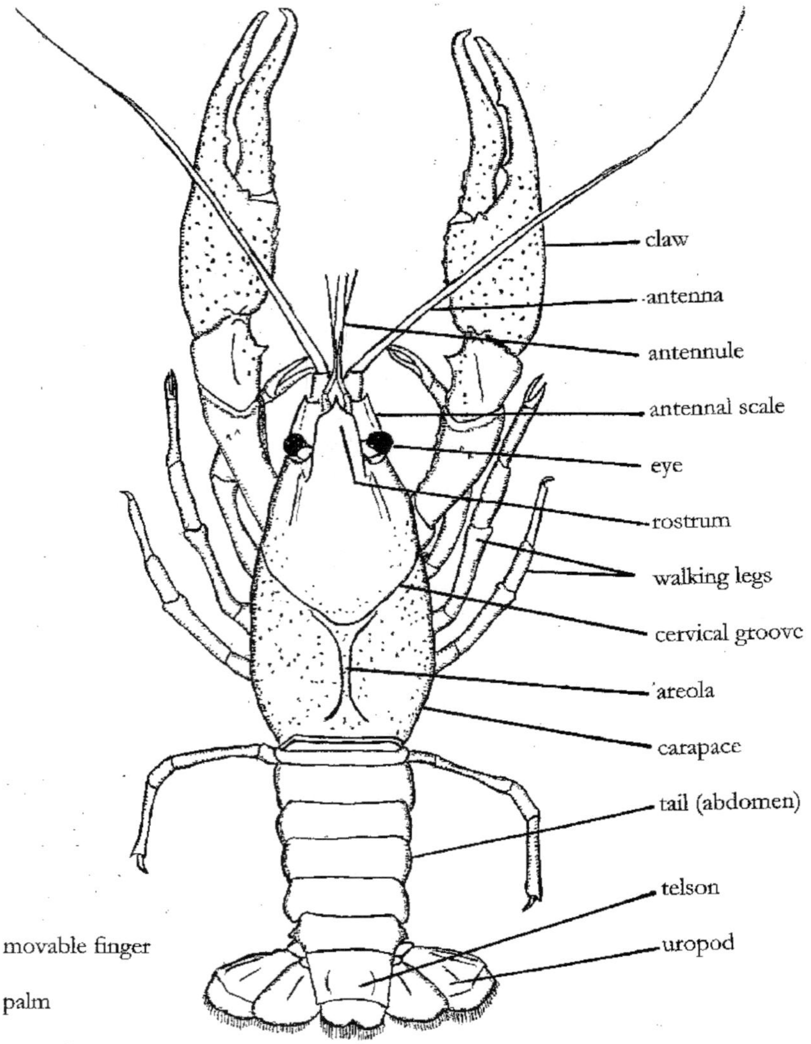

Top view of male with labeled structures.

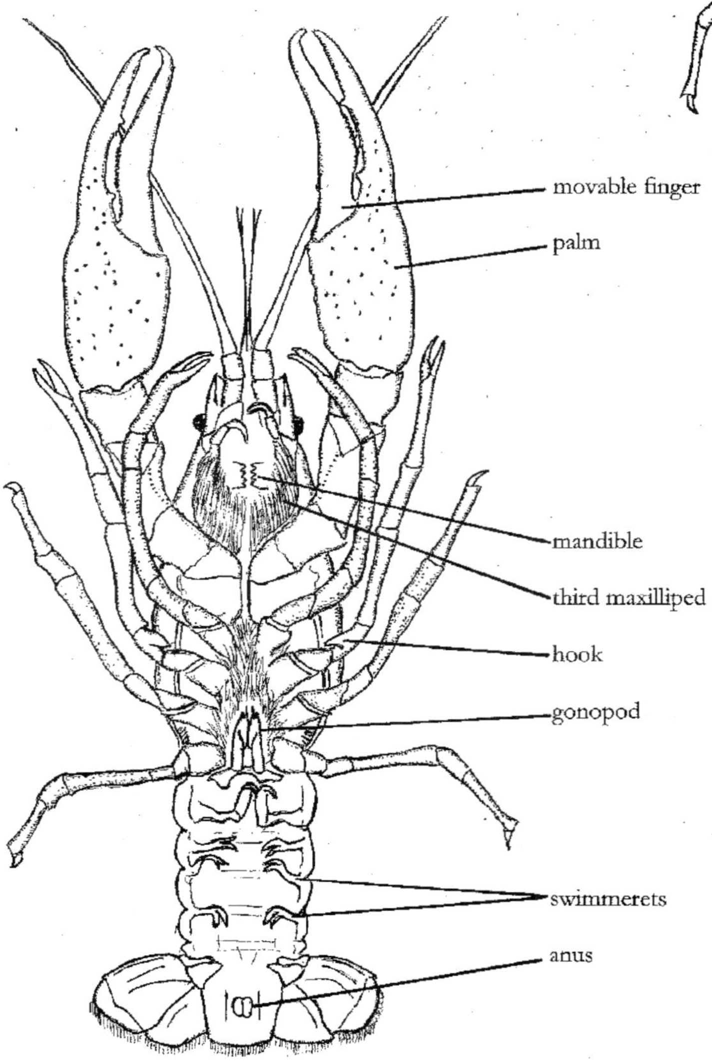

Bottom view of male with labeled structures.

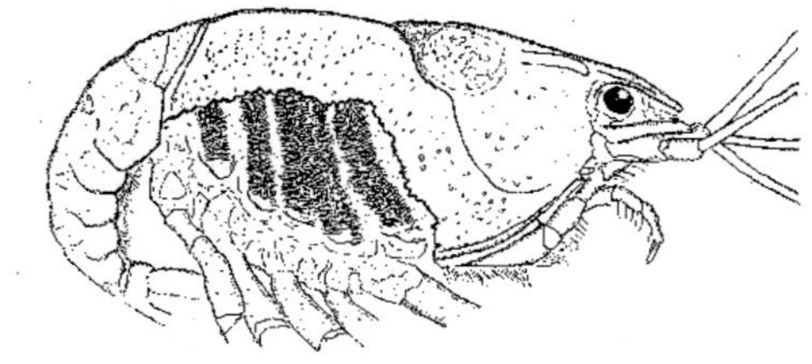

Side view (part of legs and claw not shown) with portion of the carapace removed to reveal gills.

A closer look at the head reveals black eyes born on stalks and a pointed extension between them called the rostrum. The rostrum is flattened top to bottom and, depending on the species, may be narrow or wide. Pairs of elongated, flexible antennae and less lengthy antennules extend forward in front of the eyes. Extending forward below the antennules and eyes is a pair of antennal scales. Noticeable from above, each member narrows to a point.

The sides of the carapace's thorax cover the gills. Each covering conceals the gills within a chamber that opens below to allow water to enter and cover the gills. Flattened appendages near the mouth move repeatedly when under water. Movements of these "bailers" produce a water flow that enters each gill chamber from below. The water passes over the gills and exits the chambers near their front. The flow of fresh water brings dissolved oxygen to the gills for extraction and carries away dissolved wastes released by the gills. Oxygen need not be extracted from a flooded chamber. If the gills are moist, oxygen can also transfer from air.

On top of the carapace and over the thorax portion, a pair of curved grooves marks the upper limits of the gill chambers. The area between them is called the areola.

A series of hard parts covers the top of the tail. Flexibility of the joints between each hard part and the tail's underside makes tail movement possible. The tail ends as a fan composed of a terminal segment called the telson and side appendages called uropods.

Spines and tubercles may be seen on body surfaces. Spines are hard, pointed outgrowths that protrude from carapace sides and appendage joints. Tubercles, are lower, knob-like protuberances that rise from the exoskeleton's surface, commonly the claws. Examples of tubercles are the teeth-like protuberances on the inner margin of claw fingers and the small projecting bumps on the top of the claw's palm.

Tiny pits or depressions may be present on some exoskeleton surfaces. Called punctae, they are particularly noticeable on the palm and carapace of some crawfish.

Bottom view

Looking at a specimen from below, one finds the mouth between two serrated edges of the mandible. To its side and layered backward to the claws are five pairs of appendages used for food manipulation and water movement. Their ends are directed forward and the biggest pair, the third maxilliped, now seen as on top in bottom view, is the most obvious. Also apparent is where the walking legs connect to the thorax and the division between the head/ thorax (cephalothorax) and tail. Several pairs of small swimmerets can be seen extending upward from the flexible and transparent covering of the tail muscles. In females, the swimmerets are of similar size except the first pair, which is reduced. The female annulus ventralis, also known as the sperm receptacle, is easily located between the last pair of walking legs.

In males, the first two swimmerete pairs develop differently and are larger than the others. They are important in mating. The term "gonopod" is used for members of the first pair. The gonopods are thick and stiffened. From their connection at the front underside of the tail, the gonopods extend forward and are normally held tight against the body between the bases of the legs.

The bottom of the tail is covered with a thin and transparent sternum through which may be seen the tail muscles. The end of the digestive system, the anus, is found on the bottom side of the telson at its middle.

Fine hairlike bristles are on the fringes of hard parts, ends of appendages, between leg bases, between the fingers and on the inner margin of the claw. These are called setae and are more prominent on mature animals. Tail fringes of breeding stage females usually support a heavy presence of setae.

Male or female?

The most obvious way to determine if a Texas crawfish is male or female is to look for the presence of a sperm receptacle (annulus ventralis) or a pair of gonopods on the underside of the body. The difference is easy to see in both young and old animals. Once one becomes familiar with a particular crawfish variety, other features such as tail width and claw size are helpful in sorting males from females.

Male and female crawfish with labeled structures.

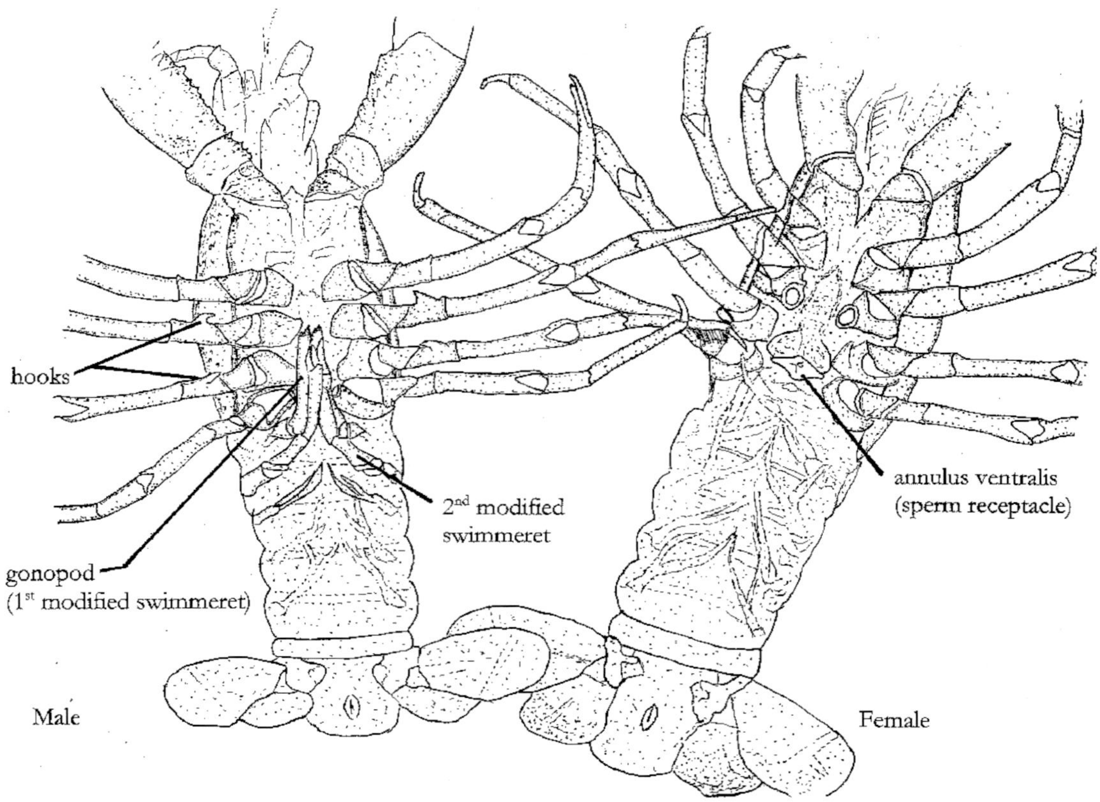

Color

Crawfish exhibit a great variety of colors and color patterns. Typically, they are lighter on the underside than on the top of their body and appendages. Crawfishes that spend substantial time above ground usually have a color pattern on the upper, exposed body surface that blends into the surroundings. Primary burrowers and subterranean species, in contrast, usually show less of a dark, blending pattern or sometime show little pigmentation at all. Many burrowers have bright, showy coloration even as juveniles.

Areas of dark pigmentation (markings) have various names. Large dark areas that contrast with lighter background are given names such as stripes, saddles and girdles. Small dark areas are called specks (very small), spots (somewhat larger) and blotches (larger with an irregular margin). Stripes are variously mentioned as dark or light, depending on perspective.

Handling live crawdads

Live crawfish are best held by one of two ways; handling them otherwise will result in getting pinched by a claw. Both methods use the thumb and index finger, or, if the animal is large, the thumb, index finger and middle finger. The first way to hold the crawfish is from behind and above with the thumb and finger placed on each side of the carapace at its mid length. By grasping it in this manner, the crawfish cannot lift or bend the claws back far enough to pinch its holder. This handling method is suitable for grabbing an agitated crawfish that holds both claws up poised for attack. Touching the underside of the head with a stick entices the claws to grasp forward and downward clearing the way for a grab with the fingers.

The second way to hold the crawfish is from in front and above, grasping the animal in a way that holds the claws forward and together while the fingers and thumb clamp the sides of the carapace toward its front. Held properly, the tips of the claw's fingers do not come in contact with the hand. If one tries to hold a crawfish by the claws only, it will attempt to break free by rejecting a claw or claws. If this happens, the specimen is much less desirable. The second method is best if one intends to measure total length. Held in this way, the tail is likely to remain straight for the measurement.

Sometimes, crawdads pinch nets and other objects and refuse to let go. To encourage them to release, first dangle them above a flat surface and then slowly lower them until the tail and walking legs touch the surface. If this does not work, lower them into a container that holds a small amount of water. While in the water, they almost always turn loose.

Two ways of holding a crawdad.

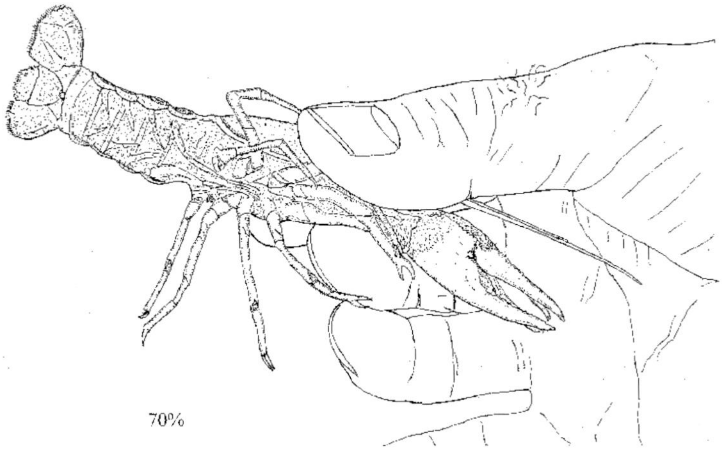

70%

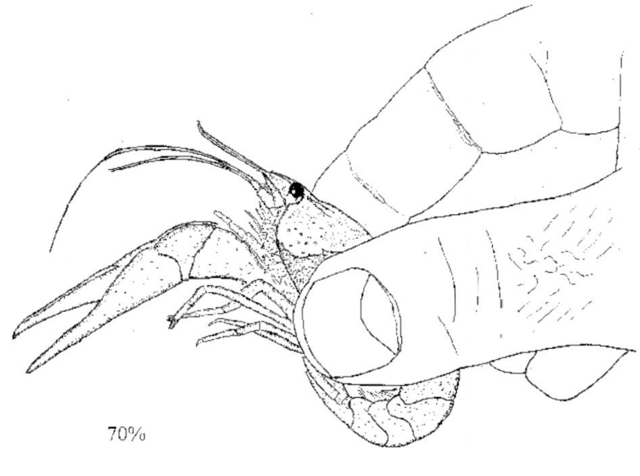

70%

Where Crawdads Live

Crawfish are freshwater animals. Many kinds spend their lives entirely in water that stands or flows above ground. Others live entirely in the water or moist conditions of underground spaces. Most kinds are able to live in both.

Habitat categories

Open waters also called surface water habitats, are either permanent (always or nearly always present) or temporary (not present in dry periods). Examples of permanent open water are streams, ponds, lakes and canals. Temporary open water is seasonally present in the case of larger ditches, wet weather creeks and wetlands, or briefly present in the case of shallow ditches and poorly drained lowlands. The latter may retain water cover for only the few days following rains.

Burrow habitats are built by the crawfish themselves, either into dry land or open water bottoms. Burrows are the typical underground spaces of crawfishes. The general habitat includes the burrow and small area that surrounds its opening.

Subterranean habitats are underground channels and caves containing water suitable for animal life. Examples are streams that course beneath the land surface, underground spaces formed by old flow or land movement that are now accessible through caves or sinks and cavities within or near openings of springs. Only a few crawfish species have adapted to an exclusive life in these conditions. None are known to live in Texas. True cave species have the same general structure of other crawfishes except for their reduced eyes and white body coloration.

Surface dwelling (surface water and burrows) crawfish are occasionally found in some Texas caves. A few of these crawfish living in deeper reaches of the caverns have a noticeable degree of pigment loss.

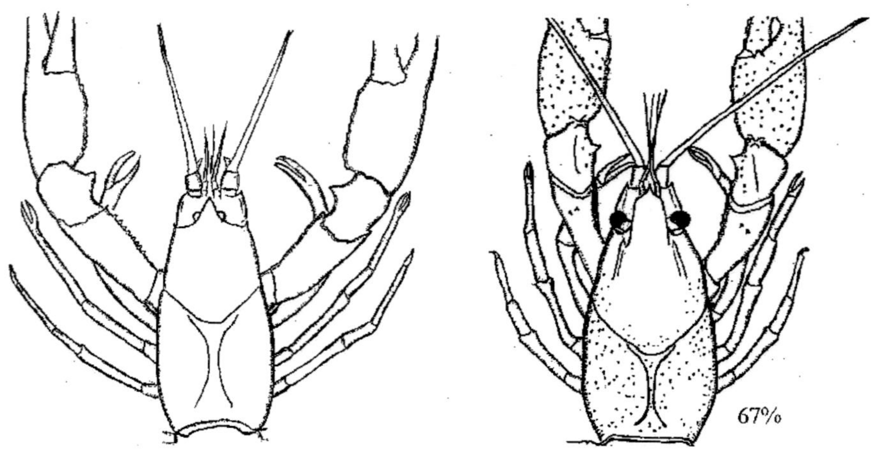

Examples of subterranean and surface dwelling crawfishes. Left: *Cambarus setosus*, a crawfish that lives in caves of southeastern Missouri and Northwestern Oklahoma. The eyes are small and not pigmented. The body is also without pigmentation (white). Right: *Procambarus simulans,* a widespread western burrower. The eyes are of average size and black. Most body pigmentation is located on the top side.

Habitat examples. Top two: Roadside ditch and ditch at culvert. Second two: Pond shoreline and Hill Country stream (note submerged rocks). Third two: Roadside backwater and deepwater creek. Forth row: Ditch in a city park.

Land burrows

When dug into dry land, tunnels usually proceed to a depth where the water table is encountered. The water table is the top of underground soil where such soil is saturated with water. As a hole goes deeper from the surface opening, its walls become moist. At some depth, the water table is reached and water seeps into the tunnel and fills the tunnel cavity.

Above the water table lies partially saturated soil capable of soaking up water brought by rains or flooding. New water may saturate the lower level of this layer causing a rise in the water table. Conversely, the water table lowers in dry times as the depth of partially saturated soil increases. A crawfish may dig deeper as the water table drops to maintain a water contact. Otherwise, a crawfish may secure a high humidity chamber until wetter times arrive by sealing the tunnel near its end or at the opening. In shallow burrows typical of some smaller crawfish species, the inhabitants form a mud casing around themselves to wait out a drought.

When water is poured into a burrow that opens onto dry ground, it eventually overflows at the surface. The water level in the hole then slowly drops as water soaks into the partially saturated soil of the tunnel sides. Similarly, a burrow that lies at the bottom of a dry ditch will fill in times of rain. When the ditch water rises and submerges the burrow opening, the burrow water becomes confluent with the new surface water. As the ditch water flows away, slowly undergoes evaporation and is absorbed by the soil, the opening is again exposed to the air. The water level in the hole then drops slowly downward.

A burrow entrance at ground level may be open to the air or plugged with a deposit of subsurface excavation. During dry periods, most, or all, burrow entrances are sealed. When rainy periods come again, plugs soften and the burrows are opened, especially the main tunnel, as excavations begin anew.

The different color of soils dug from tunnels and deposited at the surface usually contrasts with the surrounding topsoil. This makes the spotting of burrows easy. A spread or a low ridge of excavation is present around the entrances built by most burrowers. Some burrows are adorned with obvious chimneys that stand several inches above ground level. The chimney wall is made of a buildup of round pellets deposited one at a time by the crawfish. Rain, flooding and disturbances wear down most chimneys until they are rebuilt.

Crawfish species which spend most or all of their lives within and in the immediate vicinity of their burrows (primary burrowers) usually build complex lairs. A main channel typically extends downward in a spiral to the water table. From this main stem, side branches radiate outward. These then proceed downward or usually upward to the land surface. Downward ones, and some upward ones, go a short distance and dead end. The main channel extends deeper than the others, sometimes to a depth of many feet. Some species have been found to build a main channel to a depth of more than 10 feet, but most burrows go downward to a distance of less than an arm's length.

Burrows built by those species that live above ground in water provided by wet times, and below ground during dry times, are usually simple. The tunnels have an almost straight path and proceed downward in a near vertical direction. Depth usually ranges from several inches to 3 feet. One, or rarely two, side channels may open to the surface.

Burrows that open on the top of stream banks usually have side branches that open below water into the stream. These openings are along the sloping bank, or at the stream bottom.

Underwater burrows

When dug beneath permanent surface water, tunnels are usually of varying depths. Being under water, the entrances are below the level of the water table.

Simple burrows are found along the edges of lakes and ponds. Entrances usually open below water at the water's edge or out to a distance of about 2 feet. From there, the inhabitants exit to forage in nearby shoreline vegetation. If the water falls during dry periods, these entrances may be plugged. When the entrances are submerged again, the plugs are removed for access to open water.

Creeks have similar tunnels directed into their steep banks. These submerged tunnels are used for daytime shelters and for brooding and hiding. When waters fall low enough to expose many openings to the air, clay banks have a Swiss cheese look. The plugging of exposed tunnels along streams is not critical because most burrows have supplemental openings near stream bottom and because the water table is shallow along sides of permanent water creeks.

Burrows found at the bottom or sides of semipermanent streams open as simple holes from under structures such as rocks and logs, or merely from the clay substrate.

← Burrow chimneys in a dry ditch.

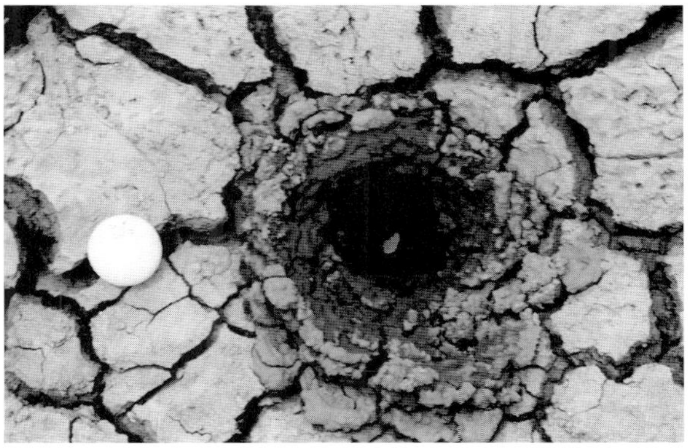

Open burrow from above. A golf ball is placed for size comparison. →

← Underwater burrows.

Plugged burrow from above. →

Crawdad Life

Burrow life

All crawfishes are able to burrow, though some, such as those that live in flowing water, do so only in times of drought, or for temporary protection. Those called primary burrowers stay below ground but emerge in times of wet weather and during humid nights. Less reclusive burrowers live in simpler tunnels of wet areas. There, they are more active above ground as they take advantage of periodic surface water.

Crawfish begin the construction of a new burrow by thrusting their claw tips into the soil. They then pull the claws laterally, spreading loose soil to the side. This is done repeatedly while the walking legs drive the crawfish headfirst into the new hole. Once the hole is large enough to contain the crawfish, it reverses the position and begins to dig tail first. The walking legs loosen soil particles. The third maxillipeds next scoop up a load of soil particles. The claws press down on the load and the crawfish carries the load up the tunnel between the claws and third maxillipeds to the surface, where the pellet is deposited. The soil removal continues in this manner until the burrow is complete. (Hobbs, 1981; Rogers and Huner, 1985).

Burrows are important to crawfish reproduction. Females of many species lay eggs and nurse young in a burrow. When rains and flooding come, the young exit the burrow and find food for growth in open water.

Life in open water

Crawfishes that live in permanent streams build simple cavities under rocks, limbs or other structures of a size that meets their fancy. They also seek out and crawl into crevices and spaces between rocks that are suitable for their size. Sites of these daytime homes are many times revealed by exposed claws or protruding fingers. Some choose a well-seated, permanent rock for a burrow cover.

Little is known about crawfishes that live in deep rivers. Usual capture methods are not effective and access to areas of deeper rock cover is possible only in times of extreme low water. Similarly, crawfish are known to occur in deep areas of reservoirs but little is known of their lives there.

Ditches, ponds and overflow pools are common habitats for crawfishes. There, larger animals usually spend days in burrows and nights in open water. Juveniles busily eat around the clock in open water. Where enough water muddiness or heavy growth of aquatic vegetation is present to protect from predators, larger crawfish search for food and build underwater burrows during daytime. Rainy or cloudy weather may also provide a stimulus for such activity.

Food

Crawfish eat plant and animal food such as small fish, insects, submerged vegetation, other crawfish, microbes attached to surfaces of rocks and plants, and dead plants with microbes growing within. Their mouthparts and pincers equip them to shred, filter, grasp and tear many food types. A large component of what they eat could be described as aquatic compost. The food intake of species living mostly in burrows is less understood than that of other types. Presumably, they consume living and dead material around the entrance to their burrow openings and underground soil within the burrows.

Nourishment

Food enters at the mouth and then passes through a short esophagus, a stomach and an intestine. Solid waste exits at the end of the tail. The intestine, called the "vein" by human consumers, lies along the midline under the top side of the hard shell that covers the tail. After food is shredded by the mandibles and then by hard parts in the stomach, a portion of the digestive fluid flows into a digestive gland and then returns to the digestive canal by the same route. The digestive gland, called the "fat" by human consumers, is a secretory and absorptive organ that fills a good portion of the thorax. Two yellow-orange lobes extend backward on either side of the digestive tract where their protrusions may be seen when the tail is removed. A yellowish color indicates that it retains a good amount of nutriment storage. From the digestive gland and intestine, processed nutrients are absorbed and enter the colorless blood.

Blood (also called hemolymph) circulation is powered by a translucent heart that lies beneath the top of the carapace near the rear. Blood passes to all internal organs and tissues. The flow travels through the gills where essential oxygen and dissolved ions are absorbed through the thin gill walls. Detrimental excess of carbon dioxide, dissolved ions and metabolic waste is released. Other waste and excess water are removed by the green glands, a pair of excretory glands plus bladders, located next to the bases of the antennae.

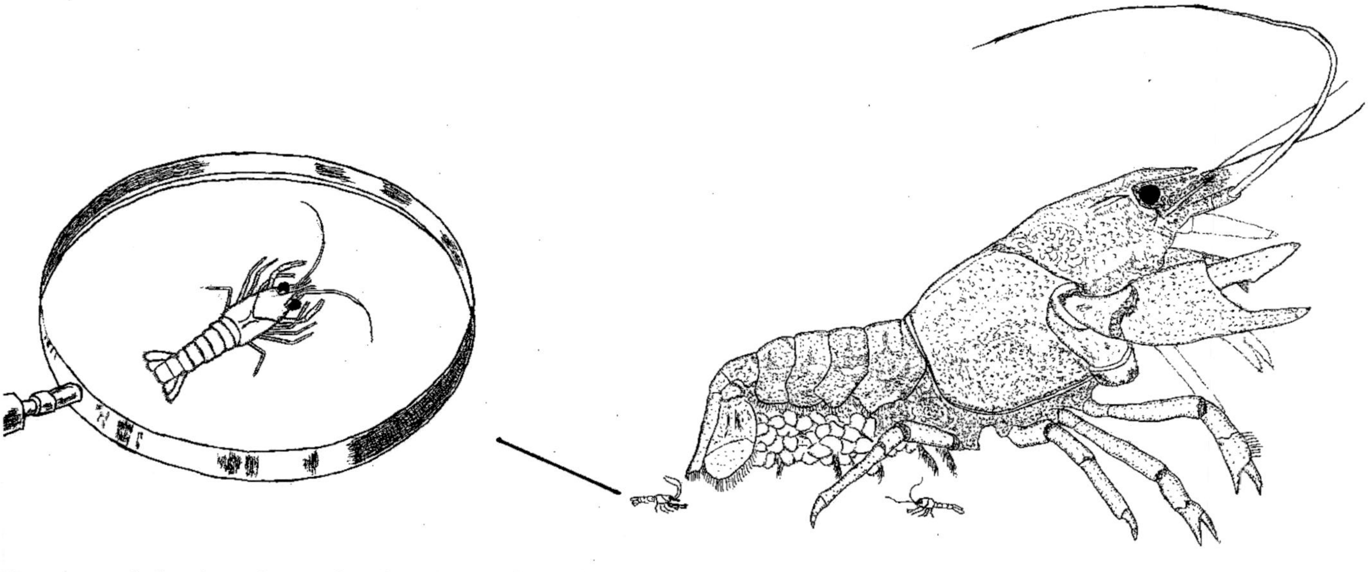

Young crawdads release from a female to begin life on their own.

Life course and reproduction

When a young crawfish departs from its mother and emerges from a burrow, a hole beneath a rock or merely releases into open water, it looks like a small version of the larger crawfish. It is ready to eat and grow. The young are voracious feeders and most species become small adults during their first year. Growth is fastest in younger individuals, and size at any particular age is largely dependent on food availability.

As they grow, the rigid shell or exoskeleton is shed at intervals. The process is called molting. Among its other benefits, the new, initially flexible shell produced at molting, favors the increase of body size.

Rapid growth occurs during early life. Young crawfish molt more than ten times during the first year and two or three times in subsequent years.

The crawfish exits its old shell in only a few minutes through a break in the back between the tail (abdomen) and carapace (cephalothorax). As the crawfish wriggles loose, the new carapace, tail, appendages, eyes and stomach lining release from the old shell. Once free, the animal looks much like it did before the emergence began.

Newly molted animals are soft to the touch. They must be handled carefully or damage could occur. These are the "soft-shelled crawfish" that are sold as delicacies in food markets and restaurants. Wild, soft-shelled individuals are vulnerable to attack by predators, including other crawfish.

Prior to shedding the rigid exoskeleton, a new, soft shell forms beneath the harder old one. Some calcium, needed to produce hardness in the new shell, is extracted from the old shell and stored in the tissues of the body. During the interval between molts, two hard stones develop between the inner lining of the stomach and adjacent tissue. When molting begins, the stones, which are composed largely of calcium, release into the new stomach and are dissolved by stomach juices. Within the next few days, the shell hardens, thanks greatly to the calcium provided from the stones and body storage. In the day after a molt, crawfish often consume their shed exoskeleton, another good source of needed calcium.

Within the size range of adults, the largest individual is often twice that of the smallest. As adults grow, claws, especially those of males, continue to enlarge proportionally to the rest of the body. Gonopods showing distinctive terminal elements with amber coloration give evidence that the age of transformation to adult — those capable of reproduction — has taken place in a male specimen. Such a male will convert to a non-breeding form (gonopods without amber coloration) with its next molt and then in its subsequent molt, back again to a breeding form with the distinctive gonopod structure. This change in male forms occurs at each molt throughout the crawfish's life. The breeding form is usually designated as Form I and the non-breeding form as Form II. The alternation in form is subject to occasional exception. For example, males that regenerate claws at a particular molting sometimes do not convert from non-breeding to breeding form.

Schematic of molting.

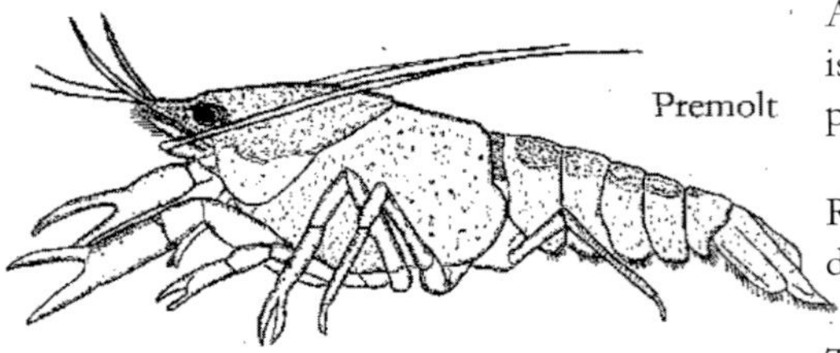

Premolt

▼

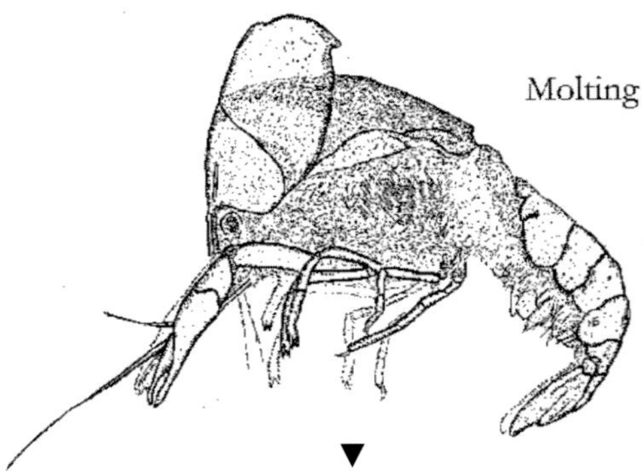

Molting

▼

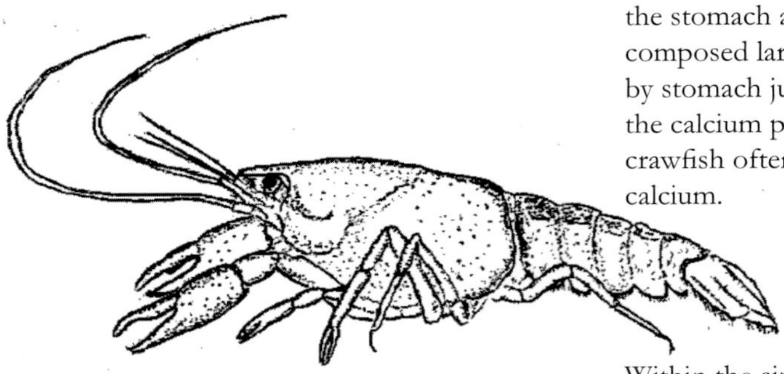

Postmolt

Shed exoskeleton

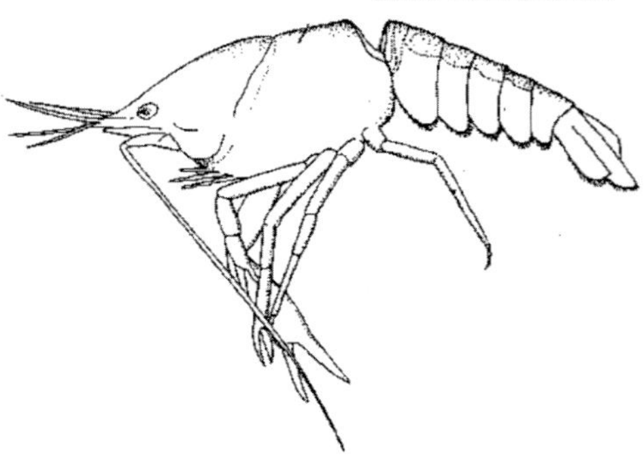

When an old adult nears life's end, the exoskeleton is thick and hard. Stream species live about three years but burrowers may attain an age of seven or eight years. Some think that subterranean species have a much longer life span.

Mating involves a breeding-form male approaching a mature female from the side and turning the female over on its back. In the process, the male's claws are used to clasp the female's clawed appendages in forward position. The leg hooks help to properly position and hold the male over the opposing ventral side of the female. The gonopod tips of the male are pressed against the annulus ventralis of the female (sperm receptacle). The gonopods are supported in this position by the pair of modified swimmerets that lie just behind the gonopod pair (also modified swimmerets) of the male. Held together, these four modified swimmerets form a tube through which sperm passes from ducts at the base of the male's last pair of walking legs to the annulus of the female. The annulus is essentially a pouch-like structure built into the exoskeleton. It is hardened with calcium except for several days after molting. Then, it is pliable and flexible enough for the male to use the elements of the gonopod tips to expand the opening for receipt of the sperm. The mating embrace may last a few minutes or more than an hour. During this time the sperm are deposited in the receptacle and a sticky substance is applied to cover the sperm. This sperm plug remains on a female for weeks or months until eggs are released from pores located at the base of the third walking legs.

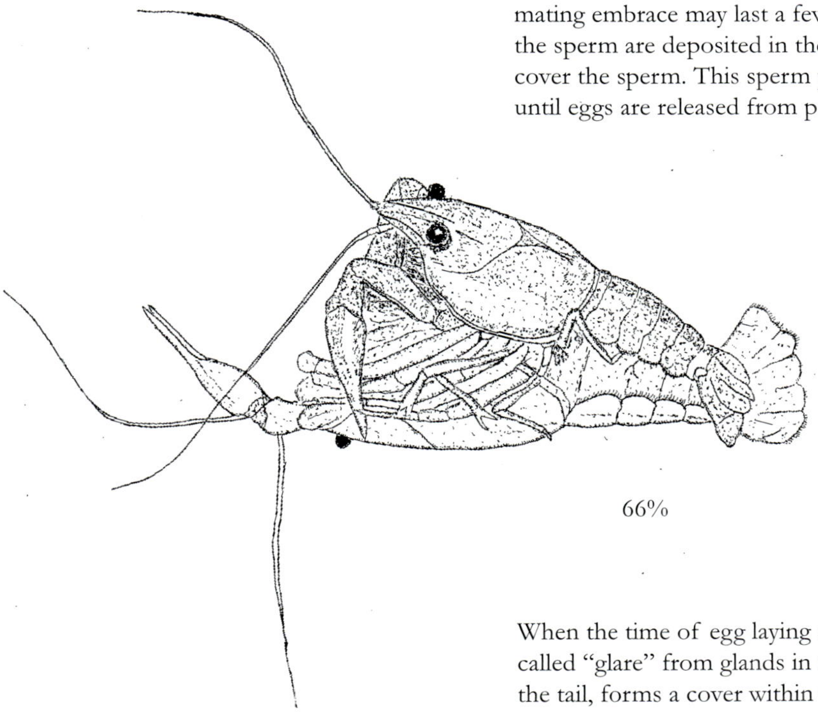

Mating embrace. The crawdad with both eyes showing is the male.

66%

When the time of egg laying arrives, the female releases a sticky substance called "glare" from glands in the tail fan. This, along with forward flexure of the tail, forms a cover within which fertilization can take place on the underside of the female's tail. Chemical substances released along with the eggs cause the sperm plug to dissolve, making stored sperm available to the eggs for fertilization. Individual eggs become attached to the swimmerets by the glare. By this time, the female has usually found seclusion in a burrow and begins an incubation period. The female keeps the eggs moist and in motion by rhythmic movements of the swimmerets. Depending on the species and stage of development, the spherical eggs have a color of black, yellow, red or brown.

After several weeks of incubation, the eggs hatch. The young remain attached to the female's swimmerets. They use yolk for nutrition at first and later consume the egg shell and incidental bits of food. Next they hold on with their claws and later, if the female is submerged, begin to release and swim back to attachment sites beneath the tail. Eventually, they release from the female altogether and begin life on their own.

Survival

Crawfish are hardy creatures relative to some others that live in fresh water. Adults, for example, are more tolerant than fish and freshwater shrimp to occasional natural degradations of water quality. Juvenile crawfish are also tolerant, though somewhat less than the adults. If oxygen becomes low in surface water or burrow water, crawfish either crawl out of that water or raise one side of the carapace above the water surface to expose moist gills to the air. High temperature of shallow, sun lit water in summer is similarly avoided by climbing up aquatic vegetation and hanging there until nighttime cooling arrives.

Burrows provide escape from both extremely high temperatures and freezing. When suitable water or humid conditions do not occur above ground, crawfish are provided a wet refuge by burrows. Dry periods normally occur in summer and early autumn in Texas though warm season rains may come at any time. In dry times, crawfish plug the burrow openings with mud. The seal conserves a humid atmosphere. In time, the plugs harden. The crawfish removes the plug when heavy rain or rising water softens it again.

Shallow underwater burrows and crevices that open in streams provide crawfish a refuge in times of swift flow. Some species seem to have a good ability to maintain position and move about in flowing water.

Perhaps the biggest threat to survival is the great array of enemies intent on using crawfish as food. Crawfish must have adequate cover to hide or they cannot withstand predation by fishes. Many would-be enemies such as birds are avoided with the protection that nighttime darkness brings. Night prowling predators like raccoons are a major threat. Raccoon footprints are a common sight along shores and in soft dirt near burrow openings. Detached claws are frequently seen in these areas indicating that a crawfish met its end or perhaps escaped by intentional sacrifice of the appendage. Crawfish are particularly vulnerable to predators and other crawfish in the period just after molting when the body surface is soft. They often retreat into hiding until the shell hardens.

Starvation may occur in crowded conditions or when food is unavailable for long periods. Crawfish have an exceptionally good nutritional reserve in their bodies, but in time even this will expire. Cannibalism often occurs in surface water pools and in burrows. The smallest crawfishes are most vulnerable.

Competitors for food and space can be of detriment to crawfish, especially an invading crawfish species that is able to out-reproduce and out-grow a local one. A local species may eventually find little to eat and become more vulnerable to its enemies.

Many crawfishes have very specific habitat needs. A lowland habitat suitable for particular species may be lost to urban expansion. A suitable stream habitat may be lost if people overuse stream water or interrupt a sufficient steam flow. Hopefully, population remnants of threatened species will survive in such circumstances until some of their members colonize a nearby agreeable habitat and recovery is made.

A need for population recovery is also occurs during times of drought. Water tables may drop 10 ft below normal during severe droughts. Many burrower populations may perish in such times. A repopulation of habitats in wetter times may result from the expansion of nearby populations that were located close to or within permanent water.

Compared to many of the larger aquatic animals, crawfish are more resistant to disease. Bacterial disease gains importance in summer when water temperature is highest in standing waters. American crawfish species are resistant to fungal infections known to be very harmful to European species. Parasites and other biological associates are found in and on crawfish but, overall, these are not very damaging. Microbes, able to affix to the shell surface, usually do not colonize to a noticeable amount unless a crawfish is in poor health or weak from starvation. Around sewerage outflows, however, the enriched water may support abundant growths of slime bacteria on shell surfaces.

Crawdad Names

Popular or common names

Unique crawdads received common or popular names as they became known by many people. Knowledgeable persons would relate or hand down the acceptable names to others and their usage became broader (part of the particular subculture).

Popular names mostly describe some distinctive appearance, behavior or place. For example, pygmy, devil, gripper, southern streamer and prairie burrower are common or popular names (English) used in this book.

A popular name may refer to a very unique crawfish or, in the case where several crawfishes look very much alike, to a small group of similar ones. Texas species in this book, for example, are discussed as one shrimp crawfish, four pygmies and two southern streamers. We chose to use certain popular names because they are widely accepted or because we heard them being used by crawfishers whom we encountered in the field.

Historically, scientific names preceded the designation of popular or common names for most crawfishes because crawfishes were not very important to or well known by people. The name crawdad or crawdaddy was good enough for most. Many crawfish species have no common or popular name.

Scientific names

While the creation of popular names for plants and animals has been going on since ancient times, science has been using a method of naming organisms, crawfish included, for several centuries. The method's intention is to develop a standardized way for people to think and communicate about the world of nature. The basic element of the method is a two-word, Latin (or Latinized) name used to designate a specific organism (species). For example, *Procambarus clarkii* is the scientific name for the red crawfish.

The first part of the two-word name indicates the genus, a group to which the species belongs. A genus represents very similar and unique kinds (or in the case of a one species assignment, one unique kind). The genus name may be used alone in formal scientific communication. The second word of the combination, called the specific epithet, is not used alone because it could be used for another animal. The combination represents the distinct species.

Genus names

At first, all known crawfishes had the same genus name. Later, recognition of common features in part of the member species led to the assignment of those crawfishes to new genera. Sometime, the discovery of a new and very different species immediately supported a new genus name. Presently, six genus names are used for Texas crawfishes: *Cambarellus*, *Cambarus*, *Fallicambarus*, *Faxonella*, *Orconectes* and *Procambarus*.

Scientific names for species

The historical naming of crawfish proceeded from one discovery to the next, here and there by different people. As one might expect, the same crawfish occasionally received different names from different scientists. When such cases are recognized, the conflict is resolved by the oldest name prevailing.

A common circumstance in the development of scientific names for crawfish occurred when several recognizable species derived from what was earlier accepted and reported as one species. The new species were named by biologists when, during continued study, they perceived features that would define separate animals. [In biological science, the eventual test that an animal represents a separate species is that it is reproductively isolated from other like animals and not just a single species exhibiting geographical variation or multiple appearance.]

Sometimes, a biologist would recognize a crawfish as distinctive but very close in appearance to another. If the biologist was not sure about its status as a new species, a variety (subspecies) designation might be made in a description of it. A distinctive Latin word would be added at the name's end to indicate the new variety. Suppose the biologist decides that the crawfish is a new variety of *Cambarus diogenes* and publishes a description with a new variety name, *Cambarus diogenes ludovicianus*. In that case, the original form is distinguished by a repeating epithet (*Cambarus diogenes diogenes*). Later, a so-named animal might be considered distinctive enough to be elevated to species status. In such cases, the subspecies word would be used as part of the new two-word name (*Cambarus ludovicianus*).

For the most part, we ignore subspecies in this book, reserving clarification of Texas subspecies to a section on scientific study.

Names are established formally in scientific literature. New species are described in "type descriptions." The description is based on representative specimens, the essential one being the type or holotype. The crawfish holotype is usually a Form I male specimen . In modern publications, a representative female is also described with a designation of allotype. In cases where the original type specimen has been lost or destroyed, a neotype is chosen. Type specimens are kept in a natural history museum for examination by present and future scientists. Type specimens of most North American crawfish reside in the Smithsonian museum in Washington, D.C.

Where some uncertainty exists about a crawfish's status as a species or subspecies, the word "form" is used in scientific communication. The word "group" usually refers to several species of very similar appearance within a genus. Subgenus is another designation scientists have used to indicate natural relationships of species within a genus.

Conversational use

People familiar with crawfishes may use common or scientific names in general conversation to describe the animals. One could use several ways to tell another person about a crawfish in hand. At first, not knowing the exact species, one might indicate the animal belongs to a particular group such as its genus or the group by which it is commonly known. For example, *Orconectes* or bluefinger. If the exact species is recognized, it could be called by its common name or the scientific name. For example, a red or *Procambarus clarkii*. More likely, in general conversation, the specific epithet (i.e., *clarkii*) would be used instead of the dual name to indicate the particular crawfish. That conversational method is often applied by people who know a group of organisms whether the assortment consists of crawfishes, fishes, bugs or trees.

In proper uses such as scientific presentations or writings, one should use the dual name fully or the abbreviated form (*P.clarkii*) to shorten text or avoid redundancy. The specific epithet is used alone (*clarkii*) when it follows mention of the dual name and there is little risk of confusion about the designation.

Crawdad Identification

General appearance

Sorting a mixed group of crawfishes into separate containers according to similar appearance is a helpful first step in identification. Next, would be the determination of the general groups to which the separate kinds belong.

Comparison of specimens to pictures in this book is useful for making tentative or presumptive identifications. Some species and species groups are unique enough for immediate recognition. Comparison of a specimen's capture location with maps or lists that indicate the varieties which occur there is also useful in identification.

Whether trying to determine the general group or a specific kind, examination of key structural features (especially those of males) is important for placement of a specimen in its proper category. For example, mature males have obvious hook-shaped structures near the base of the walking legs. The presence or absence of these structures on particular legs may be used in the assignment of specimens to their proper classification. The drawings on pages 10 and 12 show examples of third only and third and fourth leg hooks.

Though slightly variable, the areola width of male crawfish is also helpful for making identifications. If the bordering grooves of an areola are separated from one another to give a recognizable area in between, the areola is said to be "open." When the grooves are fused together, the areola is said to be "closed." An open areola may be broad or narrow depending on the crawfish group.

Different areola types.

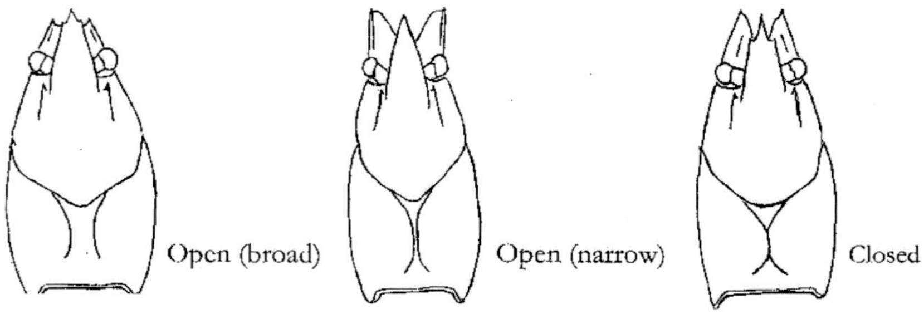

Examination of claw structure is worthwhile if one is observing claws of mature animals. Color patterns become more useful for identification as more experience is gained.

The gonopod structure of each crawfish species is unique and varies little (it is constant). The structure of this appendage is the basis of most general and specific categorization of crawfish. Examination is relatively easy. The gonopod terminates in two or several small projections. Counting the number of terminal elements and noting their general shape gives immediate recognition of the correct group to which a specimen belongs.

Confirmation of species identification requires a careful examination of a gonopod of a breeding-form male. Determining the particular kind of crawfish is more tedious than grouping, particularly if one is trying to determine identity of a crawfish that has several close relatives of similar appearance.

Examples of gonopod endings (tips) of breeding-form males. a. *Procambarus acutus*, a slenderclaw. Mesial view. Four terminal elements may be seen and also setae arising from opposite side. b. *Orconectes palmeri,* a bluefinger. Mesial view. Note two long elements. c. *Faxonella clypeata*, a gripper. Lateral view. Note one long and one short element. d. *Cambarus diogenes*, a fatclaw. Lateral view. Note strongly curved elements. e. *Procambarus dupratzi*, a southern streamer. Lateral view. Some parts of the gonopod tip are covered by setae. f. *Procambarus simulans*, a gray. Lateral view. The arrangement of terminal elements is similar for grays, hairclaws, and prairie crawfishes. g. *Procambarus clarkii*, the red crawfish. Lateral view. The large tomahawk-shaped element is distinctive. h. *Orconectes lancifer*, the shrimp crawfish. Mesial view. Note two short elements. i. *Cambarellus ninae*, a pygmy. Lateral view. Note three elongated elements. j. *Procambarus brazoriensis*, a longpalm. Lateral view. Note small, short appendages.

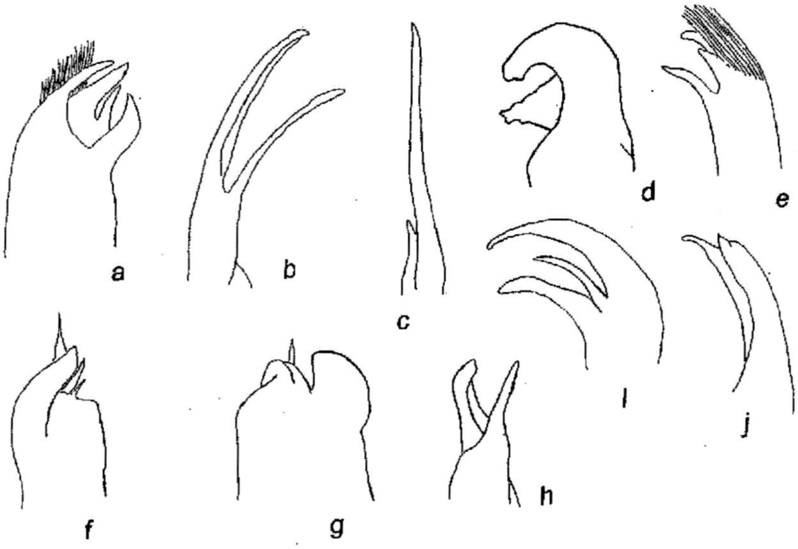

Gonopod examination

Gonopods are usually examined from a defined perspective (the direction of view). The view is established as if the gonopod were dangling straight down from an animal that is held parallel with the floor and perpendicular to the holder's eyes. The *lateral* view is the side view, one in which the outer portion of the nearest gonopod is observed. The *mesial* view is established similarly, by viewing the inner portion of the farthest gonopod as it would dangle straight down.

Because the gonopods are tucked tight against the body between the walking legs, one must carefully lift a gonopod away from the body in order to obtain a lateral or mesial view. A sharp point, such as a toothpick, works well.

When examining a gonopod with the intent of assigning the crawfish to its proper *group*, the number and general shape of the terminal elements is determined. A hand lens that gives 10X magnification or better is very helpful in making the determination.

When the gonopod is used to distinguish a *specific kind* of crawfish from close relatives of similar appearance, the best approach is to remove a gonopod from a dead, preserved, breeding-form male specimen, and examine it with a lens or dissection scope. The procedure begins by removing the appendage with forceps and placing it in a dish. Next, one establishes the proper orientation by taking advantage of the flat surface that is on the inner or opposing side of the gonopod. To obtain the lateral view, one presses the flat side of the appendage onto the bottom of the dish while viewing it with a lens or dissection scope. The mesial view is obtained by manipulating the gonopod so that the flat side faces upward toward the viewer. Orientation must be done carefully. Only a small rotation may cause the view of the gonopod and its variously directed elements to not be truly lateral or mesial and, consequently, to have a false or confusing look.

Most crawfish identification guides present figures of the female annulus ventralis. We have chosen not to emphasize this structural part as an aid to species identification. Novices would find its examination too difficult because of strong similarity between closely related species and variations that may result from size and breeding form (Wetzel, 2002).

Variety of appearance

While examination of key parts, comparison to pictures and use of range limitations give confidence to identifications, one should remember that differences, sometime remarkable, occur between *individuals of the same kind of crawfish*. A general knowledge of these expected differences helps one to avoid mistakes in identification when examining two unlike specimens of the same species.

Structural differences

Crawfish, like other animals such as ourselves, change in appearance as they grow. Perhaps the most striking changes seen during a life course are the enlargement of claws and modifications of color. Transformation in shape or proportion is more evident in some species than others. For example, claw shape changes greatly as some kinds of crawfish grow larger, while in others, the claws of little ones are very similar to those of the adults.

Differences between males and females become more pronounced with age and maturity. Male crawfish typically have larger claw size than females and possess gonopods. As they mature, claws enlarge more in males than females and hooks develop on legs. Females possess an annulus ventralis (sperm receptacle) and usually have a broader tail than males. In species whose males have a closed areola, females often have a slightly open one.

A change of form occurs when adults of most species molt. Male form, usually called Form I or Form II, alternates with each molt and is particularly apparent in the structure of the gonopod. One or more of the terminal elements of a gonopod of Form I males will be corneous (hardened) and yellowish or amber-like in color. All the elements of Form II males, in contrast, are less developed and lack the amber-like coloration.

Claws, legs, antenna and other appendages regenerate into modified replicas of the former if they are broken off. As the crawfish grows older and passes through successive molts, the regenerated parts approach the structure of the original. Regenerated claws are particularly obvious. A modified claw is usually smaller and has a different shape than an original claw. Sometime both claws are regenerated and the difference is not so apparent.

Another condition is seen rarely among crawfishes. An individual is occasionally seen to possess both male and female structures (gonopods plus a sperm receptacle).

Color differences

A crawfish species has a typical color pattern (or patterns) determined by its genetics. Slight variations are noticeable within the range of some species. Small pattern differences distinguish individuals. Differences, for example, can be seen between siblings in the layout of spots on a carapace or a claw. Color differences between sexes are usually minor. Otherwise, individual changes in color manifest with aging, molting stage and life conditions.

Pigmentation present in the skin that underlies the hard exoskeleton (the shell) is mostly responsible for the color that we see. Visible molecules (pigments) are concentrated in particular areas for pattern and appear as different colors according to their dynamic combinations.

Some crawfish species exhibit two distinctive color patterns. Both may be common throughout a species' range and even present among the individuals that live at a single locality. In other cases, individuals show a distinctive pattern in one part of the species range and not the other.

Some color variability is brought about by the life experience of the animal. Color differences are noticed when pigment disperses or contracts in color cells beneath the shell as a temporary adaptation to light exposure and background. Nutrition also has a role. Providing food that lacks certain essential components to captive animals will cause a color change.

Crawfish periodically shed the exoskeleton (molting) as part of normal growth and development. As molting approaches, the inner layer of the shell slowly disintegrates and the shell detaches from underlying tissue. The loss in normal integrity of pigment molecules results in uncommon shades of coloration that are especially noticeable in the claws. During the first day or so following molting, crawfish colors are in best appearance and may be particularly bright. As the crawfish continues life toward its next molt, the exoskeleton gradually darkens from stains and surface deposits.

Though an animal's color pattern remains the same, the intensity of color may increase or decrease as it ages and becomes sexually mature. Examples are shown later in photos of different Texas species.

Rarely, color variants are seen that are strikingly different from a species' normal color. Among these color oddities, specimens with pale blue and brilliant blue coloration are the most common. Less common are black-eyed gold or white variants. The white crawfish are a bright white like the normal condition of some cave species.

Examples of structural and color differences are shown on pages 94 and 95. The images of the following species guide also illustrate a number of the variations mentioned above. For example, specimens that show a regenerated claw may be found on pages 51 (middle image), 55 (bottom), 66 (top), 75 (top left), 77, and 82 (bottom right). The top of page 53 shows a specimen with regenerated claws.

A Guide to Texas Crawdads

All Texas crawdads belong to the same family, the Cambaridae. These crawfishes have well-developed gonopods in males and sperm receptacles in females. Some cambarids are burrowers, others flowing or standing water dwellers. Many use a mix of above and underground habitats. Some Texas species are exclusive to the state whereas the distribution of others extends into nearby states or throughout much of the continent.

For presentation in subsequent pages, the following common names are used to group the Texas crawfishes.

Pygmies
Devil fatclaw
Digger fatclaws
Grippers
Shrimp crawfish
Bluefingers
Virile crawfishes
Longpalms
Slenderclaws
Red crawfish
Southern streamers
Grays
Creekbank crawfish
Hairclaws
Navasota crawfish
Prairie burrowers

Presently, six genus names are used for Texas crawfishes. They, plus the common names for groups associated with them, are listed below:

Genus	Common name of group(s)
Cambarellus	the pygmies
Cambarus	fatclaws
Fallicambarus	fatclaws
Faxonella	grippers
Orconectes	bluefingers, shrimp crawfish and virile crawfishes
Procambarus	longpalms, slenderclaws, red crawfish, southern streamers, grays, creekbank crawfish, hairclaw, Navasota crawfish and prairie burrowers.

Where more than one species comprise a group, an introduction is given to the group as well as to the various species.

Distribution of Texas species is portrayed by mapping of county records. The maps include records of earlier scientific writings and those of our collections. Distribution in Texas river basins is explained the species descriptions. River basin presence is more definable for some species and species groups than for others. Also, a species could be found only in the upper or lower part of a basin.

Unless indicated otherwise, crawfish images are actual size. Length and width are written in inches. A conversion factor (multiply by 2.5) may be used to convert a measurement given in inches to an approximate measurement in centimeters. Drawings of the endings of Form I male gonopods are shown as an aid to species identification. They are not drawn to scale.

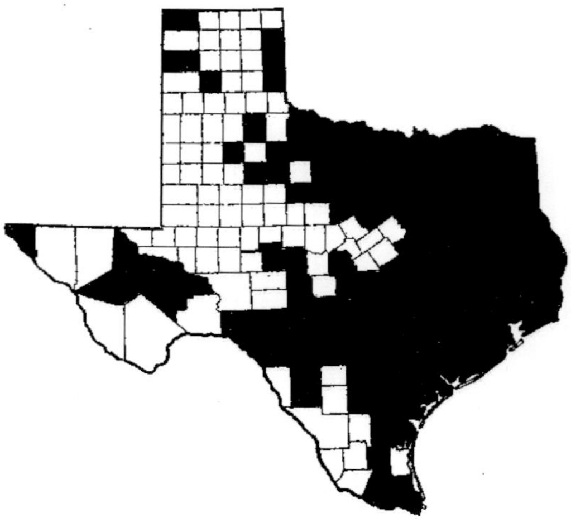

Texas counties with records of crawfish.

Cambarellus - The Pygmies

Pygmy crawfish, also called dwarf crawfish, are the smallest of all crawfishes. Length from point of the rostrum to the end of the tail of a mature male is commonly 0.8 - 1.0 inch; that of a female is commonly 1.0 to 1.2 inch. Eight species are known from the U.S. and nine others from Mexico. Texas has four.

Recognition

When sorting through a catch of very small crawfishes, the adult features of pygmies make them stand out. Usually obvious are the inflated palms of males that become more robust with age. The female tail is broader than the carapace and a few females may be carrying eggs. When pygmy males are alert or moving, the claws are held forward but with palms facing inward. This gives the observer a clear view of palm inflation when the animal is viewed from above. A broad areola is characteristic of both sexes.

The mature males of the pygmy group have hooks on the second joints of the second and third leg pair (claws counted as first leg pair). Males also possess a gonopod that terminates in three obvious processes (the terminal elements). The gross appearance of pygmy species is essentially the same. Species determination is made by examination of the structure of the gonopod of a Form I male.

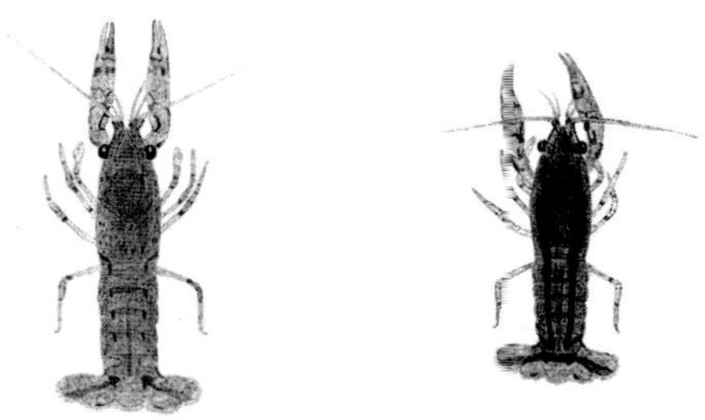

Two color patterns of *Cambarellus puer* (adult males, actual size).

Two color morphs (patterns) are known for the two most broadly ranging pygmies, *puer* and *shufeldtii.* Both color morphs may sometimes be seen for a particular species at the same locality. From above, one of the morphs shows two dark stripes, extending parallel from behind the eyes to the pre-fan segment of the tail. The other morph has two rows of obvious dark blotches that begin at the back of the carapace and proceed toward the tail fan. Animals of either pattern have a slightly bowed, dark stripe that spans the tail side-to-side. This stripe is often more apparent in females than males.

Habitat

Pygmy species are common in suitable habitats within their range. In upland or hilly regions, they may be found near streams in shallow overflow pools and ditches. In ponds and wetlands, they live in shallow water beneath protective vegetation. They are especially abundant in the flat coastal zone of Louisiana and Texas where an abundance of freshwater marshes and irrigation diversions provide a favorable habitat. There, the periodic drying of these shallow waters reduces fish numbers and thereby gives pygmies a survival advantage. More eastward in Florida, pygmies are often found in clear streams and springs. Some Mexican species are found in lakes, but others, like the U.S. species, are found in temporary, shallow waters of less than 1 foot deep. Texas pygmies pass through droughts in "cells" that they form in bottom mud. Their life span is about one and a half years.

Geographic distribution of *Cambarellus.*

Distribution

The U.S. pygmy distribution includes the coastal plain from south Texas to the Suwannee River drainage of Florida, and up the alluvial plain of the lower Mississippi River to Illinois. Other distribution in North America includes several isolated spots in Northern Mexico, the Valley of Mexico and centrally along the Pacific Slope.

Texas pygmies (*Cambarellus*)

Texas has four pygmies:

Cambarellus
ninae
puer
shufeldtii
texanus

Common names of Texas pygmy species:
The only species with established common names is *shufeldtii*. Its names include dwarf crawfish and cajun dwarf crawfish.

Distinguishing the Texas species

The Texas pygmies are distinguished by examining the gonopod of Form I males. Assumptions may be made about species identity by comparing a specimen's collection locality to the known range of each pygmy species.

Habitat and distribution of Texas species

ninae	Found only in Texas in a small range near the central Texas Coast in Aransas, Calhoun, Kleberg, Nueces, Refugio and Victoria Counties and probably San Patricio County.
puer	From the lower Brazos River Basin into eastern Texas and from there into states to the east and northeast.
shufeldtii	Broadly distributed outside of Texas, this species is found occasionally in Texas and sometime with *puer* or *texanus*.
texanus	Found only in Texas in a small range near the central Texas Coast in Colorado, Fort Bend, Jackson, Lavaca, Matagorda and Waller Counties and probably Wharton County.

The habitat is essentially the same for all Texas species — shallow waters with aquatic plant cover and underground cells into which the pygmies take refuge when waters dry up in summer or during droughts.

Catching Texas pygmies

Pygmies are best caught during a wet period in shallow ditches and pools of flat terrain. These places are common in the coastal plain where flat expanses exist and within flood plains of permanent streams. Temporary water shallower than 4 inches is ideal. A dip net with mesh tight enough to retain the little crawfish is the equipment of choice.

Form I male gonopod tips (distal ends) of Texas *Cambarellus* species (lateral view).

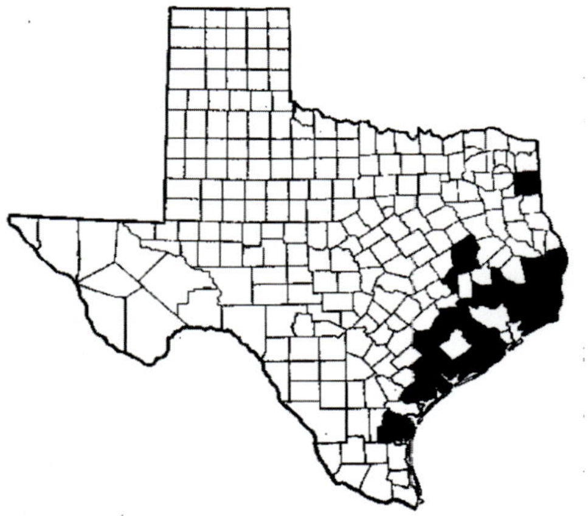

Texas counties with records of pygmy crawfishes.

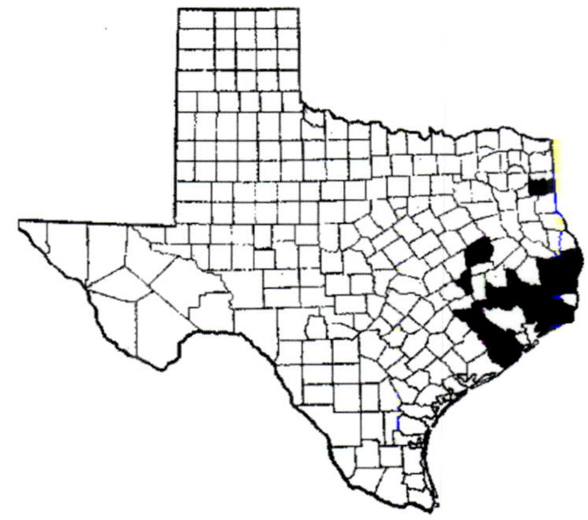

Texas counties with records of *Cambarellus puer.*

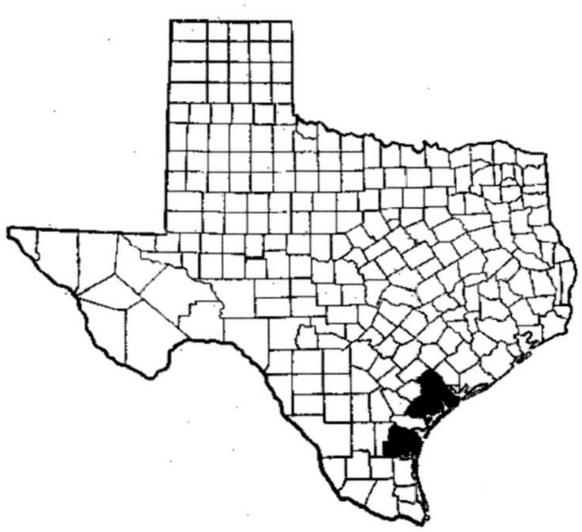

Texas counties with records of *Cambarellus ninae.*

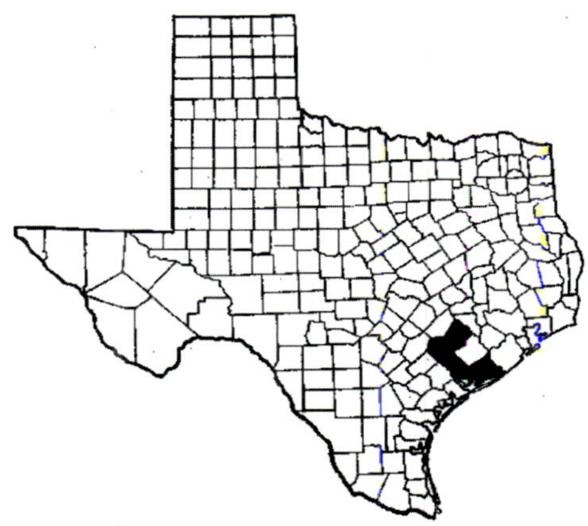

Texas counties with records of *Cambarellus texanus.*

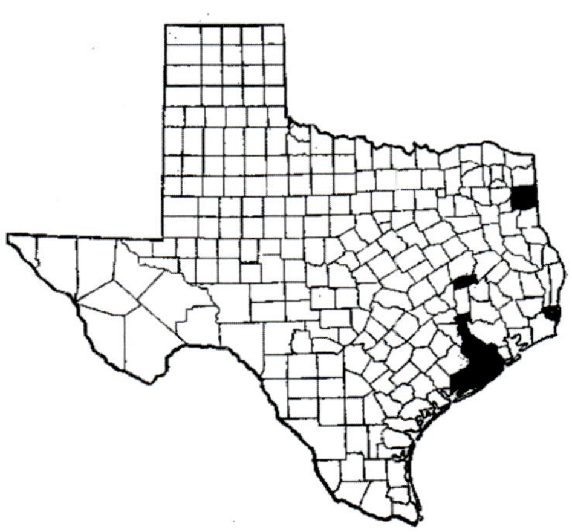

Texas counties with records of *Cambarellus shufeldtii.*

Group of *Cambarellus puer*, mostly males, Brazos County.

Cambarellus puer, female, Fort Bend County.

Group of *Cambarellus ninae*, mostly females, Kleberg County.

Cambarellus texanus males, Matagorda County.

Cambarellus texanus, female, Matagorda County.

Cambarellus shufeldtii, male, Matagorda County.

Cambarus and *Fallicambarus* - The Fatclaws

The genus *Cambarus* and genus *Fallicambarus* are together composed of 100 member species. The common name for Texas members is fatclaw. Because of their wide palms and short fingers, this type of crawfish has a perception of claw "fatness" relative to other kinds of U. S. crawfish. Fatclaws are medium (2 to 3 inches) to large (3 to 4 inches) crawfish. Some burrowers, not designated here as fatclaws, also have broad claws and short fingers, but they are seldom encountered in surface waters. The assumed life span for Texas fatclaws is four to eight years depending on species and living conditions.

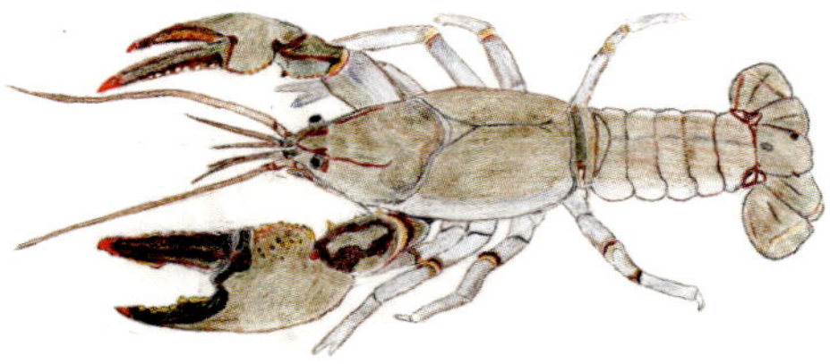

Devil fatclaw 50%

Recognition

The claws of juveniles closely resemble those of mature animals in the degree of robustness or fatness. Robust claws are common even in the smallest sizes. The palm of most species is depressed (flattened) but others have some degree of inflation. In comparison to other crawfish, most fatclaws have small eyes and, when viewed from above, a head that is blunter or less-narrowing toward its front. These features, plus a carapace that is relatively flattened on its top, give many fatclaws a stocky appearance. The distinctive characteristic of *Cambarus* and *Fallicambarus* is a male gonopod that has two, usually short and thick, terminal processes that project downward, the forward (distal) of which may curve 90 degrees or more. The unique gonopod is easily seen without magnification aids. Four of the less common *Fallicambarus* (two occurring in Texas) have a short protrusion between the bases of the two larger terminal processes.

Geographic distribution of *Cambarus* and *Fallicambarus*.

Habitat

Most fatclaws are either stream dwellers or burrowers. Several live in underground waters (cave streams). Burrowers spend all or large portions of their lives in and immediately around their burrows. Those that venture into temporary flood water return to the burrow as water recedes.

All Texas fatclaws are burrowers. The openings to their burrows may or may not have an obvious chimney. The principal tunnel extends downward 3 to 10 feet depending on water table depth.

Distribution

Fatclaws are widely distributed over North America to the west of the continental divide. The devil and digger fatclaws account for much of the broad expanse of the group's range. The majority of fatclaw species are found in the southeastern United States. Species that live in the area east of the Mississippi River are mostly (70%) stream dwellers. Except for several cave dwellers, other eastern species are burrowers. Those species west of the Mississippi are burrowers except for four that live in underground water and a couple that dwell in streams in the Ozark highlands. Texas fatclaws are burrowers.

Texas fatclaws (*Cambarus* and *Fallicambarus*)

Texas has four fatclaws:

Cambarus diogenes
Fallicambarus
    foidens
    macneesei
    devastator

Common names of Texas fatclaws: Only *diogenes* (devil crawfish) and *foidens* (digger crawfish) have well-established, common names. Other than the devil crawfish, we use the term diggers when referring to Texas fatclaws.

diogenes

foidens *macneesei* *devastator*

Form I male gonopod tips of Texas fatclaw species (lateral view).

Distinguishing Texas Fatclaws

The gonopod with recurved terminal elements easily separates the fatclaws from other crawfishes. Although each Texas fatclaw has a rather unique appearance, two species, *foidens* and *macneesei*, could possibly be confused. Both are medium size crawfishes, adults being about 2 to 3 inches in length. The other two are large species, being 2.75 to 4 inches in length.

For the less common *macneesei*, hooks are present on the third and fourth legs (claws counted as the first legs), and not only on the third legs, as is the case for *foidens* and the other Texas fatclaws. Also, the claws are more inflated than other Texas fatclaws. Two color patterns are present in *macneesei*. A stripe of lighter shading runs along the top of the carapace and tail of some individuals. In others, this stripe is not apparent and the claws are more olive green in color.

For *foidens*, the male claw is not as inflated and has a strong indentation of the movable finger on the inner margin near its base. In older "bull" males, the fingers of a claw sometimes strongly curve, the ends crossing over each other. Color pattern differences occur also in *foidens*, the light stripe on the top of the tail being narrow or broad and obvious.

The devil crawfish (*Cambarus diogenes*) has two color patterns. One is very colorful, especially in smaller individuals. The other (shown as a drawing on the previous page) is more drab except for bright red finger tips.

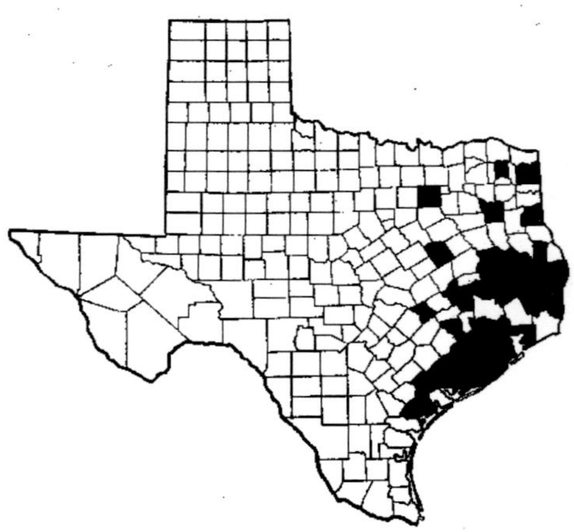

Texas counties with records of fatclaw crawfishes.

Habitat and distribution of Texas species

diogenes — The devil crawfish (*Cambarus diogenes*) burrows near permanent steams and ditches and is more catchable in surface water during spring and after rains. It usually remains within the immediate vicinity of its burrow.

foidens — Out of its burrow, the digger crawfish (*Fallicambarus foidens*) is the most likely to be found in ditches and temporarily flooded lowlands. It, like the devil, is widespread in North America.

macneesei — *Fallicambarus macneesei* is uncommon and usually stays within its burrow. During wet nights it grazes the vicinity of the burrow and sometime ventures into temporarily flooded ditches or puddles. The range of *macneesei* is the coastal plain of Texas and Louisiana, usually within 50 miles of the Gulf of Mexico.

devastator — *Fallicambarus devastator* makes elaborate burrowing systems that enlarge as dome shaped "rooms" more than 2 feet in diameter at the tunnel bottoms. "Cave ins" occur when livestock or equipment passes over the burrow fields. Like other primary burrowers, *devastator* may be caught at night when it emerges from the burrow to feed. A large chimney or "spread" usually surrounds the burrow opening. This burrower is found in woodlands and pastures of Angelina, Houston, Polk, Trinity and Tyler Counties.

Catching Texas fatclaws

These burrowing species are usually not caught with nets, except after rains in newly flooded pools and ditches. A good place to find them is in pools and ditches in low areas where terrain has little or no slope. A fatclaw is sometime discovered in a tire rut or some other small depression the day after a rainstorm. During warm humid nights of wet summers, fatclaws usually leave burrows to wander the land surface near to their burrow homes. They can be picked up by hand. In times when holes are not plugged to protect from drought, fatclaws climb to the burrow openings at night, claws and eyes protruding. There, they may be caught in the burrow opening by quickly cutting off the tunnel before they escape downward into the earth. Well-maintained schoolyards, parks and athletic fields of Texas communities in the flat coastal plain are good places to find diggers. In upland areas with more hilly topography, look for them in flat expanses near streams. The devil typically burrows into the earth near streams. Well-constructed chimneys suggest its presence. The devil will enter minnow traps set near creek banks. We caught one specimen by hand as it crawled up the bank of a small creek. Near the coast and in flat areas, the devil crawfish is commonly found at night, looking out from the burrow opening as described above. A small devil is occasionally caught with dip net from a lowland ditch.

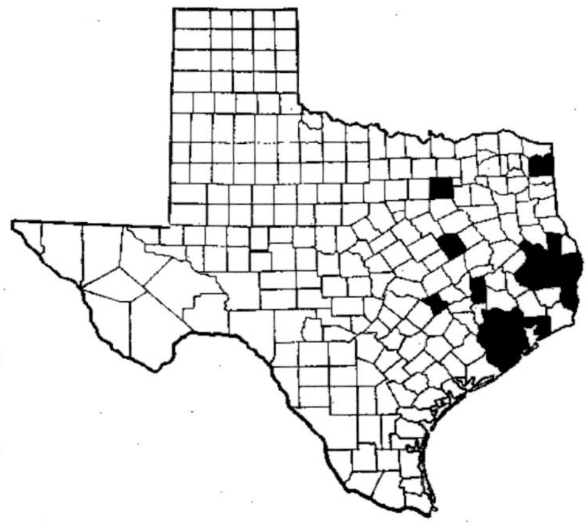

Texas counties with records of the devil crawfish, *Cambarus diogenes.*

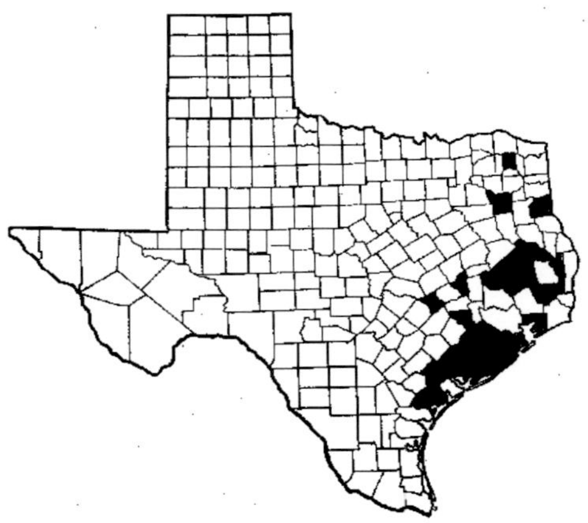

Texas counties with records of the digger crawfish, *Fallicambarus foidens.*

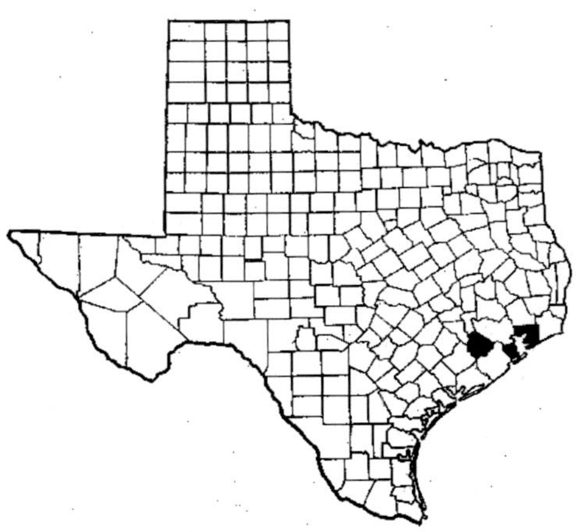

Texas counties with records of *Fallicambarus macneesei.*

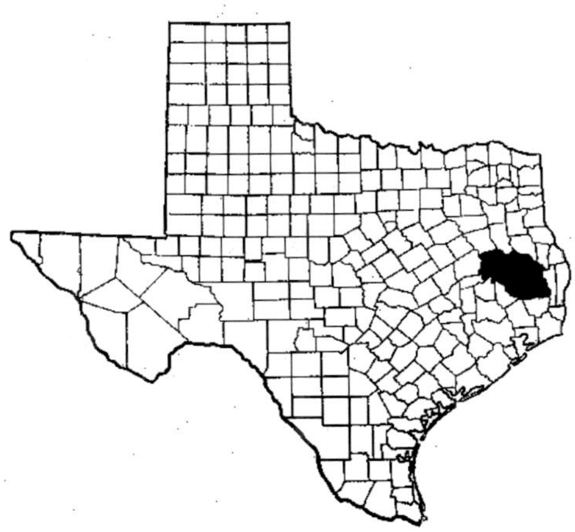

Texas counties with records of *Fallicambarus devastator.*

Examples of the devil crawfish, *Cambarus diogenes*

Male, Chambers County.

Small female, Fort Bend County.

Male, Chambers County.

Female, Chambers County.

Juvenile, Fort Bend Country.

Examples of the digger crawfish, *Fallicambarus foidens*

Male, Chambers County.

Male, Matagorda County.

Male, Chambers County.

Male, Matagorda County.

Female, Matagorda County.

Female, Hardin County.

Female, Panola County.

Examples of *Fallicambarus macneesei*

Male, Chambers County.

Male, Chambers County.

Male, Fort Bend County.

Juvenile, Chambers County.

Examples of *Fallicambarus devastator*

Female, Angelina County.

Male, Trinity County.

Female, Trinity County.

Juvenile, Angelina County.

Faxonella - The Grippers

This group of small crawfish contains four species. When handled, grippers usually grab together the ends of their opposing legs and claws as if holding an invisible tube. They maintain this posture for a while after release. Other crawfish such as the longpalms and some orcos behave similarly but to a lesser degree. Adult lengths are 1.25 to 2 inches; usual catches are 1.5 to 1.75 inch. The life span is one and a half to two years.

Geographic distribution of *Faxonella.*

Recognition

Claws of mature gripper males have inflated palms that are longer than the fingers. The palm thickness is most apparent in side view. The areola is broad in both sexes and hooks are present on the third legs of mature males. The male gonopod has two terminal processes that bend inward and overlap like a pair of crossed sabers. This crossing becomes obvious when the crawfish is viewed from below. All gripper species have the same general appearance and are best distinguished by examining their characteristic gonopod structure.

A gripper species may exhibit one or two color patterns. When a specimen is viewed from above, one pattern shows a light stripe that runs from the rostrum down the top of the tail. It is bordered by parallel dark stripes running from behind the eyes backward to the end of the tail. The other pattern has parallel dark stripes also but they are broader and usually begin at the back of the head instead of in front of the tail. The light stripe in such animals is narrower. Both patterns may appear less obvious if the darker striping shows faintly, or perhaps only on the tail.

Habitat

Grippers are usually caught within the flood plain of streams. In wet seasons they emerge from chimneyless burrows of 0.7 to 1.2 feet deep to live in shallow sloughs, swamps, ditches and overflow pools. They are best caught with dip nets from among the leaves and plants that cover the mud bottoms of these temporary waters.

Distribution

Grippers are found in the Red, Sulphur, Cypress Bayou and Sabine River Basins of eastern Texas. Above Texas, they range into southeast Oklahoma. Their range includes all of Louisiana, and, above there, southern Arkansas and the Mississippi River floodplain to southeast Missouri. From Louisiana eastward, the range extends through southern Mississippi and Alabama to the Florida panhandle. From the panhandle eastward, grippers are found in central Georgia and part of South Carolina.

Texas grippers (*Faxonella*)

Texas has three species:

Faxonella

- *beyeri*
- *blairi*
- *clypeatus*

Common names of Texas gripper species: The term mudbug has been used for the grippers and *clypeatus* in particular. Mudbug is occasionally used for crawfish in general, usually when a person is being silly or, in other situations, when the person is trying to appear savvy. The name shield crawfish has been used for *clypeatus* (means shield in Latin).

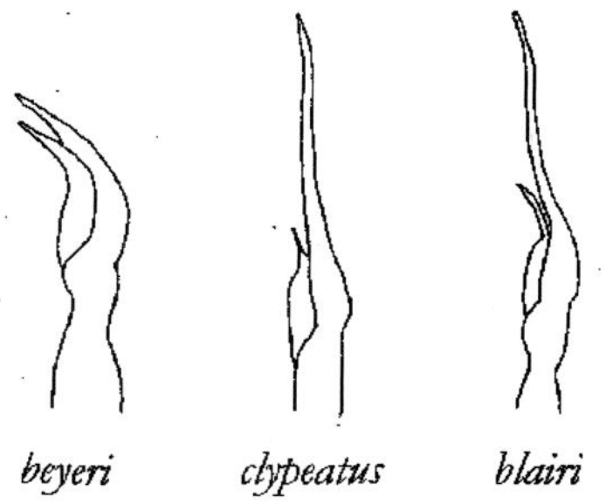

Form I gonopod tips of Texas *Faxonella* species (lateral view).

Distinguishing Texas species

Examination of the Form I male gonopod easily distinguishes the three species. The longer terminal element of *Faxonella clypeatus* is very slender and elongated. That of *Faxonella beyeri* is thicker and bends inward at a greater angle. *Faxonella blairi* has an elongated, slender terminal element like *Faxonella clypeatus*, but the basal element is larger and longer.

Distribution and habitat of Texas species

beyeri	Sabine Basin from Harrison to Newton Counties.
blairi	Red River Basin, Red River County and probably other counties of this basin in the northeast corner of the state.
clypeatus	Sulphur River and Cypress Bayou Basins (tributaries of the Red River) in northeast Texas.

All three species use the same type of habitat. The typical habitat is a shallow pool of standing water located near a permanent stream. Often, the pool lies under or near hardwood trees growing in a flatwood.

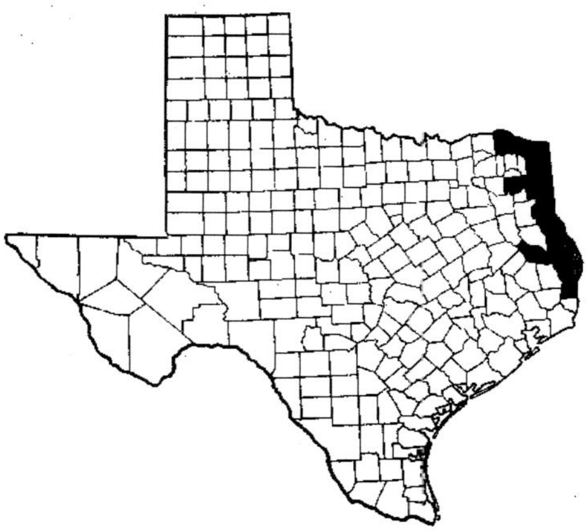

Texas counties with records of gripper crawfishes.

Catching grippers

Grippers are usually caught in overflow pools of larger creeks and rivers. Dip net and seine methods are adequate.

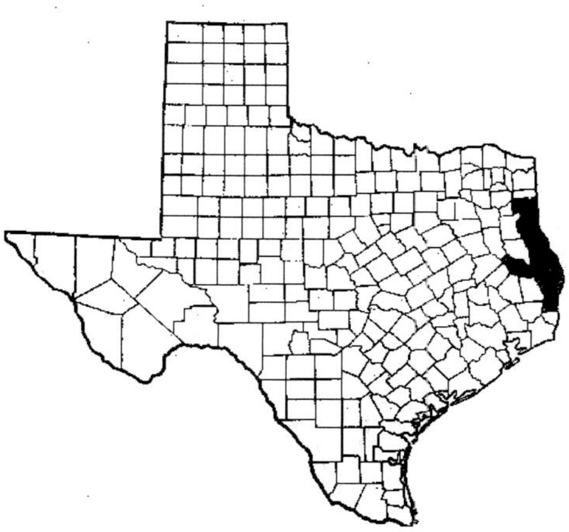

Texas counties with records of *Faxonella beyeri.*

Texas counties with records of *Faxonella clypeatus.*

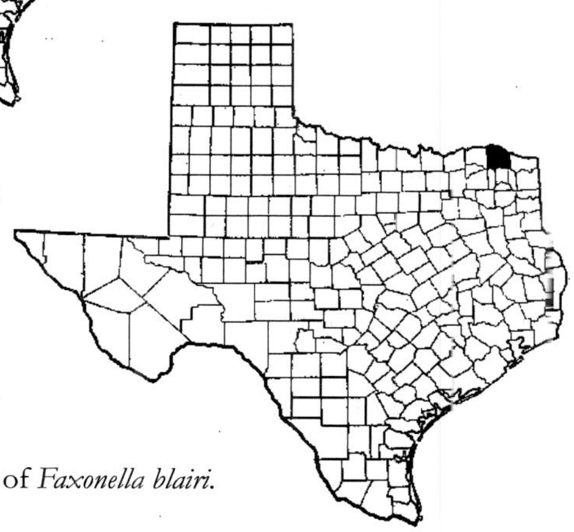

Texas counties with records of *Faxonella blairi.*

Examples of Texas grippers – *Faxonella beyeri* and *Faxonella clypeatus*

Male *beyeri*, Panola County.

Female *beyeri*, Panola County.

Female *clypeatus*, Bowie County.

Female *clypeatus*, Cass County.

Male *clypeatus*, Cass County.

Male *clypeatus* Cass County.

Male *clypeatus*, Upshur County.

Orconectes

The genus *Orconectes* is a large crawfish group of more than 75 species. Orcos are important as food for other animals and several are used widely as fish bait. Size varies according to species, adult lengths of the smaller species being 1 to 2 inches and the larger ones 4 to 5 inches. The claws of older individuals of some species grow large, the length of a palm and its fingers exceeding that of the carapace. Compared with other types of crawfishes, most larger orco species are relatively docile when handled. The smaller species are more aggressive. The life span for Texas species is not known but may be assumed to be two to four years.

Recognition

All Texas orcos have a gonopod bearing two terminal elements. Checking the male gonopod for two obvious terminal elements of similar size is an easy way to determine that a crawfish belongs to the group. Though most *Cambarus* and *Fallicambarus* also have two terminal elements, theirs are thick and usually more curved. Discernment of individual species is somewhat difficult due to close similarity of gonopod structure between species and the potential for structural variation of a particular species' gonopod.

Habitat

Orcos live primarily in streams and lakes where they use rocks, root masses, litter or other solid materials for cover during daytime. At night they emerge to become active. With the exception of some eastern mountain areas and southern river basins of the U.S., orcos are the most common crawfishes found in flowing, rocky sections of streams. Some species attain large populations in lakes and contribute to the diet of fishes. When lakes have an abundance of suitable hiding places, many individuals of a species grow old and huge. Thus, they out-grow the mouths of most of their enemies.

Distribution

Human transfers (often accidental as bait) of several species have resulted in establishment of populations beyond their native ranges. Orcos are now present in all contiguous U.S. states except Nevada and Florida and in the Canadian provinces bordering the U.S., except British Columbia and New Brunswick. Introductions have been reported for Chihuahua, Mexico (as a probable introduction) and Baja California Norte, Mexico.

Geographical distribution of *Orconectes*.

Orconectes - Shrimp Crawfish

The unique appearance of the shrimp crawfish, *Orconectes lancifer,* makes it easily distinguishable. The rostrum and antennal scales are long and pointed and the claws do not become very robust in older animals. The shrimp crawfish is found in lakes and sluggish streams. It is rarely caught. Adult length is usually 2 to 3 inches.

The known range of the shrimp crawfish is the Mississippi River drainage from southern Louisiana to southmost Illinois, including the Red River branch as far as northeast Texas. One report displayed further range from Lake Pontchartrain watershed in Louisiana across coastal Mississippi and Alabama, but gave no supportive data.

In Texas, we have found this crawfish only in a sluggish stream that courses through a woodland to the Red River in Red River County.

Geographic distribution of the shrimp crawfish, *Orconectes lancifer.*

Form I gonopod tip of the shrimp crawfish (mesial view).

Orconectes - bluefingers and virile crawfishes

Aside from the shrimp crawfish, Texas orcos are similar enough to present as one account. Within that assemblage, however, structural differences between the various accepted species are difficult to use (even by descriptive naturalists), singly or in combination, for a definitive portrayal of each kind. Thus, the Texas species and their look-alikes found elsewhere are hard to understand and explain (an enigma). A short discussion intended to provide better understanding of the group's complexity is given later in this book.

Texas bluefinger and virile crawfishes

Orco species, other than the shrimp crawfish, known to live in Texas:
Orconectes

palmeri
deanae
difficilis
virilis
nais

Common names of Texas bluefingers and virile crawfishes: Only *virilis* (the virile crawfish) has an established common name. The common name, northern crawfish, has also been used for this species.

Distinguishing Texas species

Several of the above, which we call bluefingers, have prominent blue to blue-green coloration of the fingers in juvenile and adult sizes. Blotches or flecks are distributed on the claw tops. The males have a closed areola. All are very similar in general appearance. Examination of the Form I male gonopod for curvature of the terminal elements, their separation angle and the proportion of angle length to stalk length is required for species assignment. The Texas blue fingered crawfishes are *Orconectes palmeri, Orconectes deanae* and *Oсrconectes difficilis*. The known lengths of adults are *palmeri*, 2 to 4 inches; *deanae*, 2 to more than 3 inches; and *difficilis*, 2 to more than 3.5 inches.

Three varieties of *palmeri* reside in Texas. The first, living along the Red River and sparsely in eastern Texas, is somewhat larger than the other two. The others live farther west in the Hill Country, are medium in size and are more similar to a variety that lives near the Mississippi River. They do not show obvious blue coloration of fingers in larger sizes though smaller individuals express the color. Again, gonopod structure is a good way to determine differences, even between these varieties.

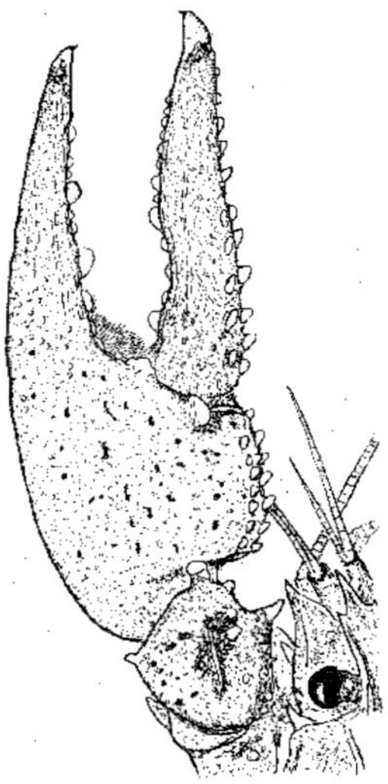

Texas *Orconectes*, except for the shrimp crawfish, have prominent whitish or yellowish tubercles along the inner margin of the movable finger and palm which sharply contrast with a darker base coloration of the claw.

The virile crawfishes represent another distinctive type of *Orconectes*. They are similar to the bluefingers but differences between the two types are apparent. The size range of adults is 2 to 5 inches. Claws enlarge with age, especially those of males, and in older animals claw length (palm and fingers) exceeds that of the carapace. The typical coloration, especially a light yellow to orange coloration of the tubercles on the inner margins of fingers and inside margins of claws, is helpful in recognizing this type of crawfish. The ground color of the upper side of carapace and abdomen is brown and the top of the claws is a grayish green. Fingers may have the same coloration as the rest of the claw or may be blue-green before the tips (bluishness less prominent than that of bluefingers). Finger tips are orange-red. The areola is somewhat narrow but open. We have collected two virile crawfishes in river basins west of the Trinity River. The two species, *virilis* and *nais*, are distinguished by gonopod structure. The mesial view of a Form I *virilis* shows the lengthier element to curve toward its end, beginning that curve beyond the fork of the two elements. In *nais*, the element curves more strongly beginning at the fork and the curve.

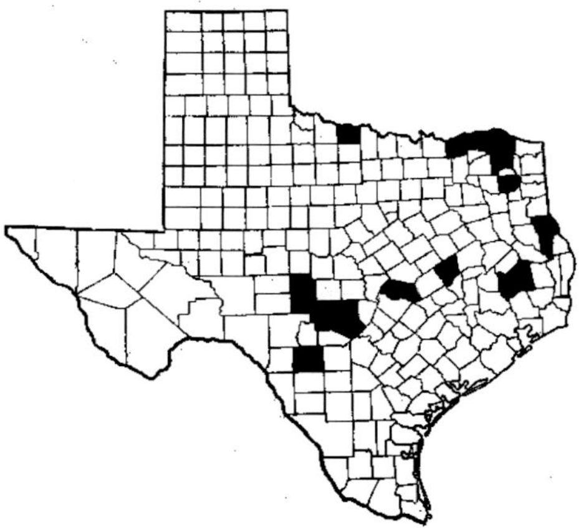

Texas counties with records of *Orconectes*.

Habitat and distribution of Texas species

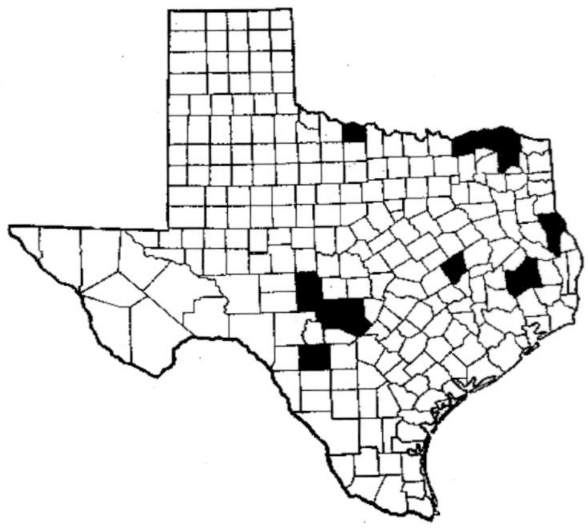

Texas counties with records of *Orconectes palmeri.*

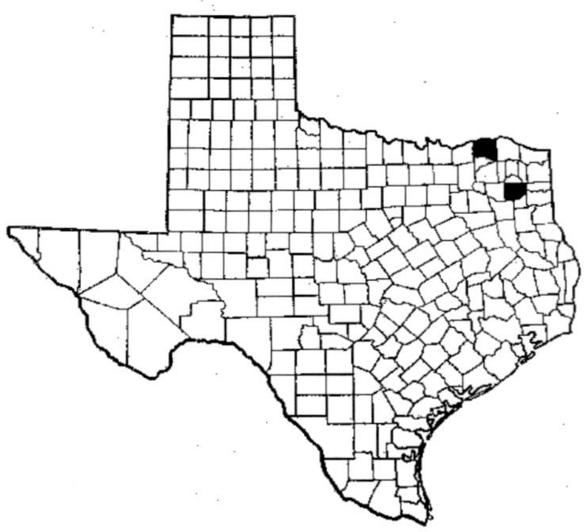

Texas counties with records of *Orconectes difficilis.*

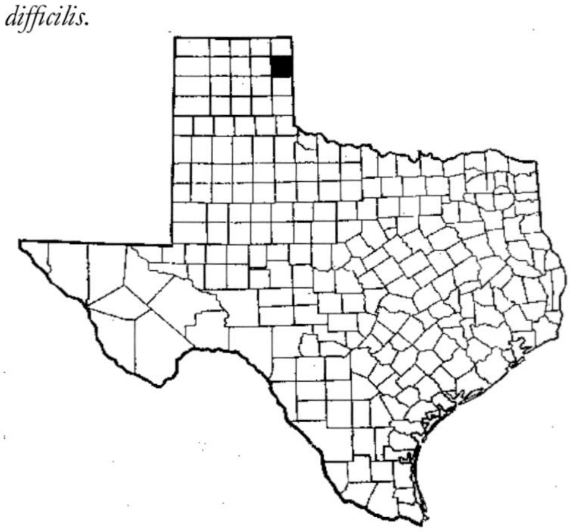

Texas counties with records of *Orconectes deanae.*

palmeri One variety of *Orconectes palmeri* is common in the Red River basin along the northern border of the state. It is occasionally caught in the eastern part of the state. Its range includes much of the central U.S. west of the Mississippi River. Holes in steep banks and root masses are the favored cover for daytime life in more permanent streams. The Red River tributaries are usually shallow except during rainy periods. Submerged rocks and logs become the favored sites for daytime refuge in the shallow stream beds.

As mentioned above, a *palmeri* variety occurs in streams close to the lower portion of the Mississippi River. A very similar variety lives in the Hill Country portion of the Guadalupe River. There it remains under cover of cypress roots and shoreline rock of the main channel only to venture out during overcast days and night. Because of periodic "gully washers," most rocks of the main channel are too well seated to support hiding places.

Another variety, almost identical to that in the Guadalupe River except for gonopod structure, occurs in the Frio River in Uvalde County. The habitat is rocky cover of shallow stream flows. There the crawfish spends its days under rocks and its nights venturing about the rocky substrate. The Frio, a tributary to the Nueces River, is a spring fed stream with a limited cover of small rocks over a solid rock base. This species uses the cover of these small rock patches where they hollow out crevices down to the hard pan.

difficilis Red River Basin. The habitat is the same as for *palmeri.*

deanae *Orconectes deanae* is found in Lake Marvin, a small lake that is located in the Texas Panhandle. Previously, it was known from localities farther up the Canadian basin in New Mexico and farther down that basin in Oklahoma. A daytime habitat is typically any available solid cover that provides a small confinement inaccessible to larger predatory fish.

virilis The range of the virile crawfish is comprised of most of the United States and Canada east of the continental divide, except for introductions in California and Arizona. The Southeast is free of them from the Carolinas to Texas, except for pockets of introduction in Alabama, Mississippi and Tennessee. Human introduction may or may not be responsible for isolated populations found in Texas and the Mexican state of Chihuahua. We have collected this species in the San Gabriel River in Williamson County, Texas. Its habitat is a flowing stream where rocks or other structures are used for daytime cover. An acceptable habitat is also slow flowing or standing water of stream impoundments and deep stream pools.

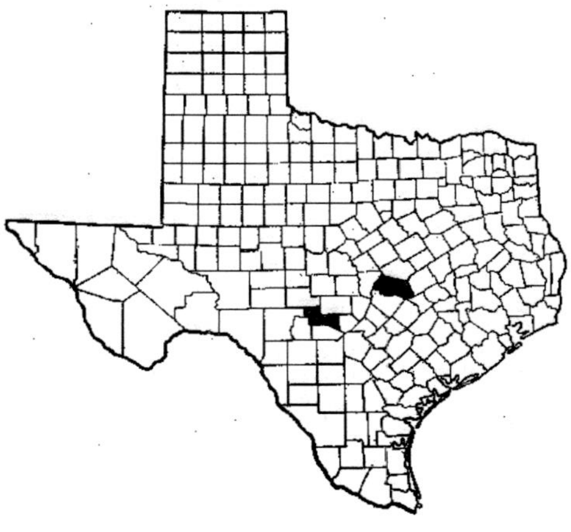

Texas counties with records of *Orconectes virilis* or *Orconectes nais.*

nais The range limits of *nais* is poorly known. Our Texas collection localities are in the Guadalupe River in Kerr County. More collection could locate *nais* elsewhere up or down the Guadalupe. We have not, however, caught it above Ingram. The habitat is a flowing stream or cool spring-fed backwater. Long pool runs produced by shallow overflow dams are also used by this crawfish. During daylight hours, *nais* may be found in stream shallows under rocks and other structures.

Catching Texas orcos

Orcos are usually caught from the clear water of flowing streams or nearby seepage pools. A good method is to hand turn rocks or debris in shallow water. Look for rocks that have "clean sweeps" on the stream bottom just around a rock's margin. Immediately after turning the cover, the newly exposed water is momentarily clouded with silt. As the water clears, the resident may be hand pinned or netted with a dip net.

Some reservoirs support orco populations. These usually have rocky bottoms or some other type of structural protection. Trapping is best in most cases but turning rocks along shallow shorelines is also useful.

In the Texas Hill Country, orco populations increase in pool stretches established by low dams that are commonly built along larger creeks and rivers. Traps may be employed.

Orcos are sometimes caught close to a permanent river in a small tributary creek. There, they may be caught from shallow pools that form during dry weather or remain after flood water recedes.

Because orcos are active at night, dip netting in well-known areas of shallow streams with a headlamp is usually a successful catch method. Rainy seasons are poor times to catch orcos because of high water. When wading low water of streams, crawfishers should be aware of the danger of water surge brought from a heavy upstream thunderstorm.

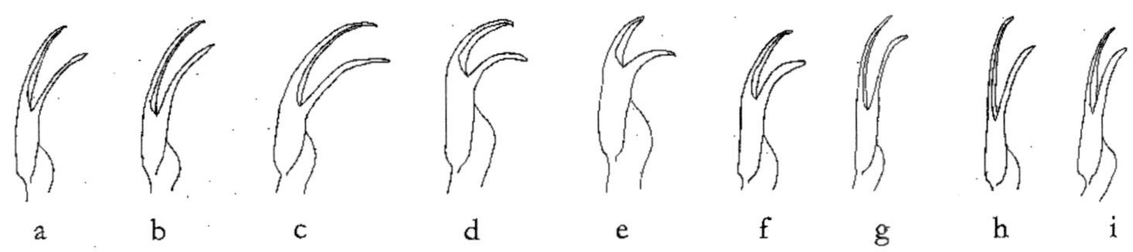

Form I gonopod tips of *Orconectes* species (mesial view). a. *palmeri*, Red River basin. b. *palmeri*, Guadalupe River. c. *palmeri*, Frio River. d. *deanae*. e. *difficilis*. f. *nais*, Guadalupe River. g. *virilis*, San Gabriel River. h. *virilis*, Colorado. i. *virilis*, Missouri.

Juvenile *lancifer*, Red River County.

Male *lancifer*, Red River County.

Male *palmeri*, Lamar County.

Male *palmeri*, Fannin County.

Female *palmeri*, Fannin County.

Male, Polk County (movable finger tips broken off).

Male, Polk County.

Male, Polk County.

Examples of *Orconectes palmeri*

Male, Guadalupe River, Kerr County.

Female, Guadalupe River, Kerr County.

Juveniles, Guadalupe River, Kerr County.

Male, Frio River, Uvalde County.

Male, Guadalupe River, Kerr County.

Male, Frio River, Uvalde County.

Female, Frio River, Uvalde County.

Male *difficilis*, Lamar County.

Male *difficilis*, Lamar County.

Female *deanae*, Hemphill County.

Female *difficilis*, Lamar County.

Male *deanae*, Hemphill County.

Male *deanae*, Hemphill County.

Male, Williamson County.

Male, Williamson County.

Female, Williamson County.

Female, Kerr County.

Male, Kerr County.

Male, Kerr County.

Procambarus

The North American genus *Procambarus* contains more than 150 species, making it the most diverse crawfish group in the world. Smaller species have adult lengths of approximately 1.5 inch whereas the larger species grow to more than 4 inches. The life span of species living continually in open water may be as little as two years whereas species spending lengthy periods in burrows may exceed six years.

Recognition

Characteristic of *Procambarus* species is a male gonopod that terminates in three or more tiny elements. These elements appear as a small cluster on the tip of the main stalk that bears them.

Geographical distribution of *Procambarus*.

General appearance may be used to assign a member of *Procambarus* into one of the several groups of closely related species. Crawfish species within such groups look very much alike. Nine groups are used to categorize Texas *Procambarus* species in this book. A summary of he distinctive features of each group is given on pages 95 and 96.

Determination of the exact species is made by examination of gonopods. Because of the small size of the elements, a dissection microscope may be needed to conduct the examination. Habitat preference, geographical distribution and lifestyle are also supportive of designating the unique members or member groups of the genus *Procambarus.*

Habitat

Some species live in ponds, lakes or streams and have burrows under bottom structures or along the banks. They enter these for protection or for survival in times of drought. Some species live under ground in wet areas that flood in times of heavy rainfall. Then they emerge from burrows and enter the new surface water to forage, find mates and undergo molting until the water disappears. Others live more or less exclusively in or near burrows built far from permanent or semipermanent surface water. They temporarily leave their tunnels during humid or rainy nights to venture outward over land for short distances.

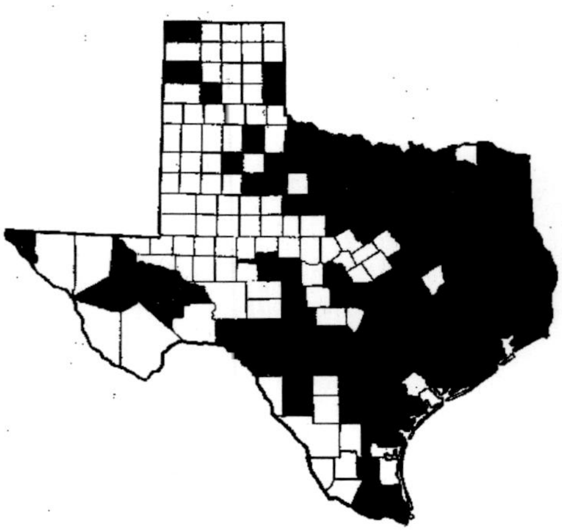

Texas counties with records of *Procambarus* species.

Distribution

Most species are found in the southeastern U.S., but the native range of the genus includes central and eastern North America. Accidental or intentional introductions by humans have expanded occurrence beyond the native range, mostly from uses as fish bait, ornamentation or aquaculture.

Procambarus - Longpalms

Longpalms are small crawfish. Adult length is 1.2 to 2 inches. The palm of the claw of the mature male is tubular in shape and long in proportion to the fingers. The palm of smaller males is not as lengthy. Females have a smaller and less-developed claw.

Recognition

Mature males have claw palms that are obviously longer than their fingers. The areola is broad. Over a tan base are dark stripes that extend from behind the eyes to the pre-fan segment of the tail. Either broad or dark striping may be found in a species. If broad, the space between them appears as a bright pinstripe. Smaller longpalms could be confused with larger striped pygmies. Unlike pygmies, longpalms lack the prominent band across the tail fan and the males have hooks on legs three and four, instead of two and three. Longpalms could be confused also with grippers of similar size. Grippers, however, do not show intense darkness of the stripes, only bear hooks on leg pair three and have long crossing gonopods.

Geographic distribution of the longpalm crawfishes.

Habitat

Longpalms are caught primarily in standing or slowly moving water such as is found in roadside ditches. Such water disappears during periods without rainfall and the longpalms take refuge in their burrows.

Distribution

The range of the group extends from southeast Texas eastward to the Mississippi River. In Louisiana, there is a northward extension west of the Mississippi River.

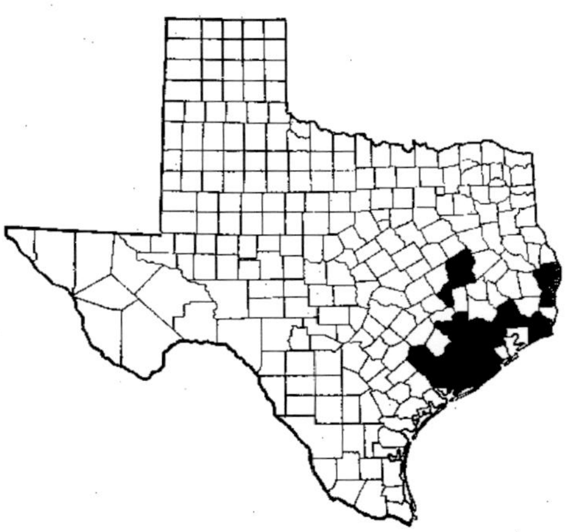

Texas counties with records of longpalm crawfishes.

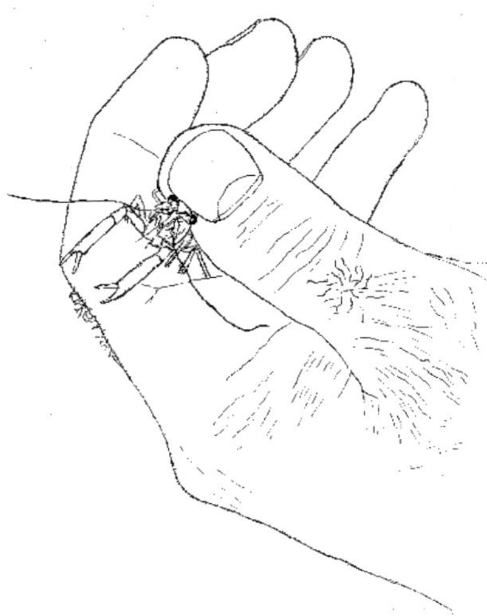

Longpalm males have claw fingers that are obviously shorter than the palm.

50 %

Texas Longpalms

All of the three longpalm species occur in Texas:

Procambarus
brazoriensis
hinei
incilis

Common names for Texas longpalms: No vernacular names are in common use for these species.

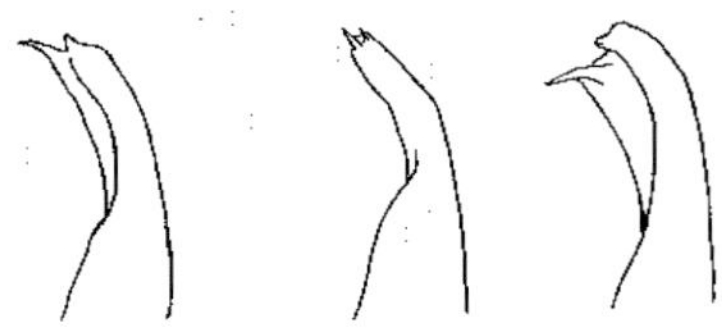

Form I male gonopod tips of Texas longpalms (lateral view).

Distinguishing Texas species

The appearance of all three species is essentially the same. The gonopod is distinctive but because the adults are of small size, the tiny gonopod is best examined using some means of magnification.

Distribution of Texas species

brazoriensis	Within 20 miles of the Gulf of Mexico in Brazoria County.
hinei	Southeastern Texas and Louisiana, generally from the Houston area to north Louisiana and from there east of the Mississippi River to the Gulf.
incilis	West and southwest of *hinei* and within 140 miles of the Gulf.

Catching Texas longpalms

Longpalms are mostly found near permanent water in ditches and overflow pools. Dip nets catch them in water shallower than 1 foot. Pygmies are often caught with, or nearby longpalms.

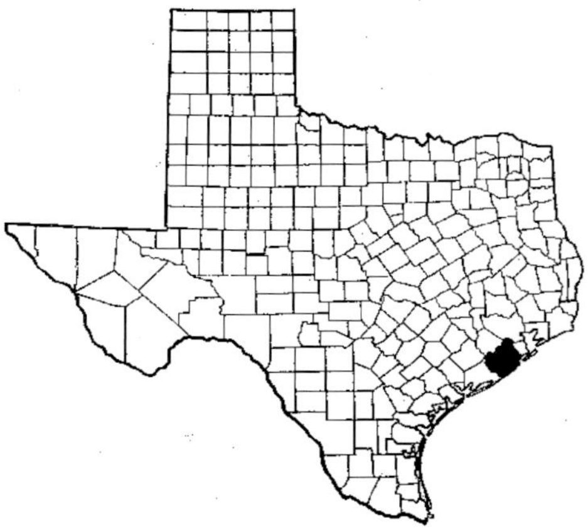

Texas counties with records of *Procambarus brazoriensis.*

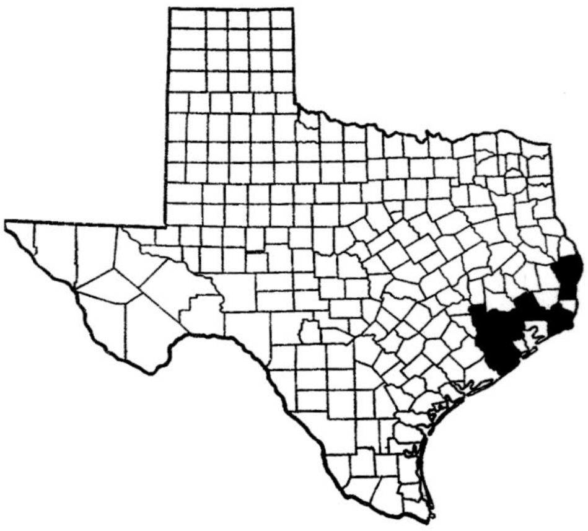

Texas counties with records of *Procambarus hinei.*

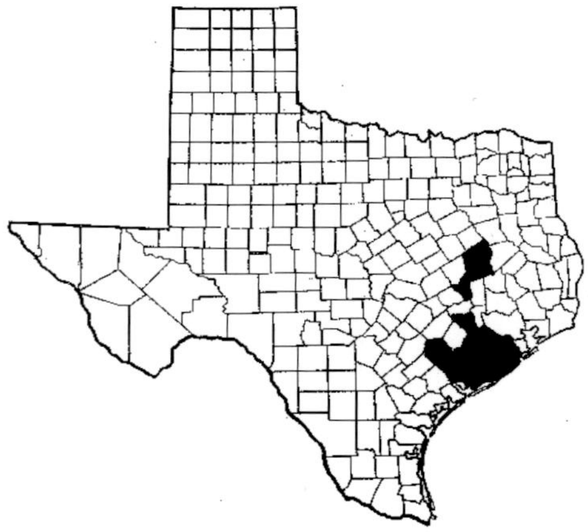

Texas counties with records of *Procambarus incilis.*

Examples of *Procambarus incilis*, *Procambarus hinei* and *Procambarus brazoriensis*

Male *incilis*, Brazos County.

Female *incilis*, Brazos County.

Male *incilis*, Matagorda County.

Female *incilis*, Brazos County.

Male and Female *hinei*, Fort Bend County.

Male *hinei*, Fort Bend County.

Male *brazoriensis*, Brazoria County.

Male *brazoriensis*, Brazoria County.

Female *brazoriensis*, Brazoria County.

Procambarus - Slenderclaws

The species of this group of large crawfishes are almost indistinguishable by general observation. The larger juvenile and small adult males have claws with narrow palms and long fingers. This gives their claws a slender appearance. As males grow to their largest size, the palm broadens and the claw loses part of the slender characteristic of younger individuals. Females have a more inflated palm and shorter fingers. One species, *Procambrus zonangulus*, is often grown along with the red crawfish in south Louisiana and Texas aquaculture operations. In more northern areas of the mid-south, *Procambarus acutus* is also used as an aquaculture species.

Recognition

A conspicuous broad dark stripe along the top of the tail is obvious in males and females. Below the stripe on each side is a narrower dark stripe that runs from in front of the tail to the back of the carapace. In some individuals, the narrow, dark stripes continue forward onto the carapace, usually stopping behind the cephalic portion. Examination of the gonopod is necessary to distinguish slenderclaw species.

The red crawfish is often caught along with slenderclaws. It is also a large crawfish with a broad tail stripe. Slenderclaws may be quickly distinguished from the red crawfish by noting that they have dark tubercles on the top of the palm and that they lack the broad, blunt terminal element characteristic of the red (see red crawfish, recognition).

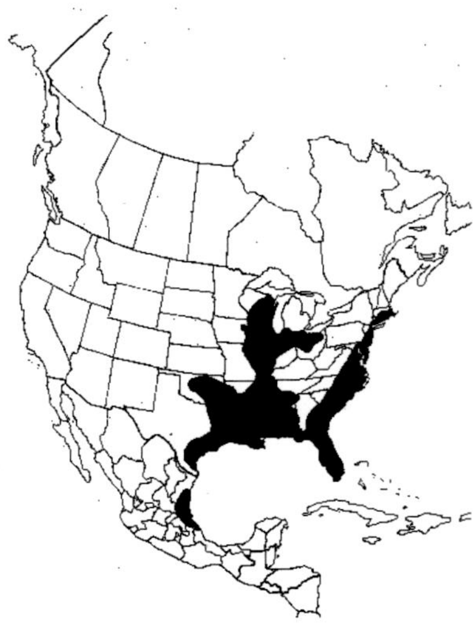

Distribution of slenderclaw crawfishes.

Habitat

Slenderclaws are very common in ponds and overflow pools near streams where they make burrows along the shoreline. They are common inhabitants of roadside ditches.

Distribution

The range of the many slenderclaw crawfishes covers much of North America. Due to human introductions, isolated populations are present elsewhere.

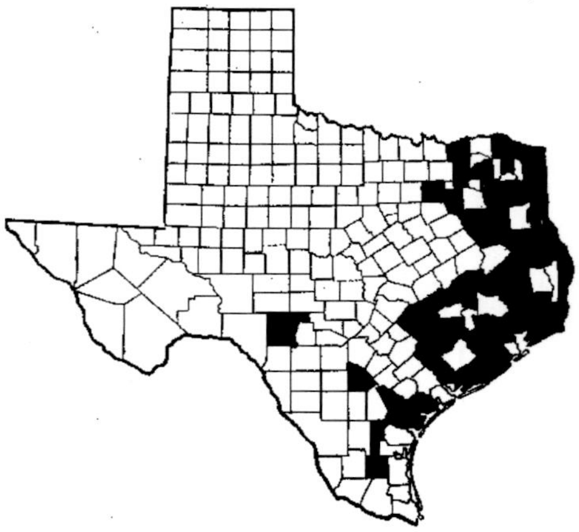

Texas counties with records of slenderclaw crawfishes.

Texas Slenderclaws

Texas has six slenderclaws:

Procambarus
- *acutus*
- *nechesae*
- *nueces*
- *texanus*
- *zonangulus*
- unnamed species

Common names of slenderclaws: Somewhat established names for *Procambarus acutus* are white river crawfish and white crawfish.

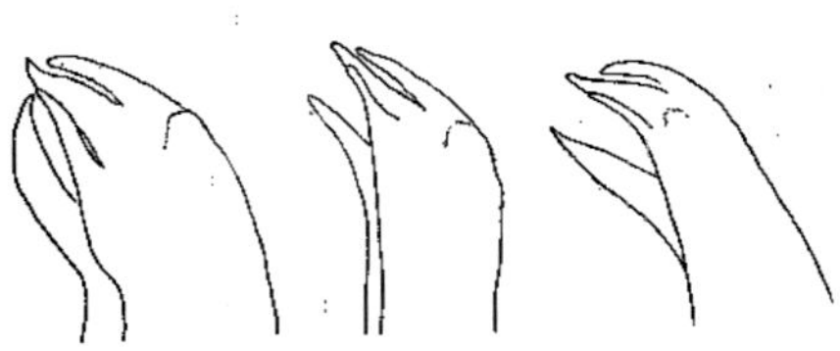

acutus *nueces* *zonangulus*

texanus *nechesae*

Form I gonopod tips of slenderclaw species from Texas (lateral view). Last tip = unnamed species of south Texas.

Distinguishing Texas species

Allowing for variation in color within a species, the general appearance of all Texas slenderclaws is basically the same. All slenderclaw crawfish of Texas were once thought to be *Procambarus acutus*. As more became known, differences were recognized (the key being gonopod structure) and additional species were established. Examination of the gonopod is necessary to be sure of the species.

Distribution of Texas slenderclaws

acutus	The range of *acutus* is throughout a large portion of North America and generally north and east of that of *zonangulus* .
nechesae	Neches River Basin, apparently in the upper portion.
nueces	Nueces River Basin.
texanus	The known distribution of this species is a very limited occurrence in Bastrop County.
zonangulus	The range of *zonangulus* is within the flat coastal plain from Texas to Louisiana. Considering the limited amount of information on the species, its range apparently ends midway along Texas' flat coastal plain. In that area, slenderclaws have features of both *zonangulus* and *texanus*.
unnamed sp.	We collected specimens of this undescribed species from temporary water habitats that lie within the watersheds of Copano (Aransas, Refugio, and San Patricio Counties) and Baffin Bays (Brooks and Jim Wells Counties). These locations are within 30 miles of coastal waters.

Maps showing county records of slenderclaw species are not presented because more study is required to establish reliable distributions of the long-known and newly-known species.

Habitat

Slenderclaws occur in a wide variety of habitats. They favor non-flowing surface water habitats such as ponds, flooded wetlands and roadside ditches. There, they live a burrow, open water existence.

Catching Texas slenderclaws

In ditches shallower than 1 foot, dip nets are useful. Seines work well in ditches of up to 2 feet deep and 5 to 8 feet wide. A good haul in such places may land many crawfish. Near to, or just within burrows that line pond shorelines, slenderclaws may be caught with swift hand grabs or with a combination of hand grabbing and tunnel blocking. Another way to catch them along shorelines is to use a rake or seine to pull ashore vegetation and then separate the crawfish from the mix.

Male *Procambarus acutus*, Harrison County.

Examples of *Procambarus acutus*

Male, Lee County.

Male, Lamar County.

Male *nechesae*, Angelina County.

Male *nechesae*, Angelina County.

Male *nechesae*, Polk County.

Male *texanus*, Bastrop County.

Female, Atascosa County.

Male, Atascosa County.

Female, Atascosa County.

Male, Jefferson County.

Male, Jefferson County.

Female, Jefferson County.

Juveniles, Jefferson County.

Male, Brazoria County.

Male, Jefferson County.

Examples of unnamed species occurring near the Gulf in South Texas

Male, Jim Wells County.

Male, Refugio County.

Female, Brooks County.

Female, Refugio County.

Male, Brooks County.

Procambarus - Red Crawfish

The Red Crawfish, *Procambarus clarkii*, is also known in some areas as Louisiana Red or Red Swamp Crawfish and to certain kids as the red snapper. The reddish coloration of adults is more intense than the redness that develops in the later life of most other American species. Unlike adults, juveniles do not have the bright reddish coloration, but their fingers are usually red.

The red crawfish grows to more than 5 inches. As a product of aquaculture and wild capture, the red crawfish is important for human food and fish bait.

Though the Red belongs to a small subgroup of *Procambarus* containing six species, it is responsible for most of the subgroup's total range and is the only species occurring in Texas. The other species are found in smaller areas in Florida, Georgia, Alabama and South Carolina. One species is found in central Mexico.

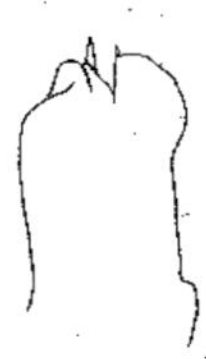

Form I gonopod tip of the red crawfish, *Procambarus clarkii* (lateral view).

Recognition

The red crawfish is rather aggressive and is able to quickly extend and lift the claws backward in a large arc toward the tail. Tubercles on the claws are cream colored in juveniles but become more reddish as the animal ages. In older adults, the base color develops as a dark purple with the surface tubercles, especially on the claws, appearing brilliant red in contrast. Mature males have a gonopod that bears a broad and blunt terminal element. This is easily recognizable with a keen eye when the gonopod is pried from its resting place with a probe or blade. A hand lens, though usually unnecessary, may be helpful. A broad stripe runs along the top of the tail of both sexes and is more obvious in lighter colored individuals. Though helpful in recognizing red crawfish, such a stripe is also present and more prominent in the slender-claw crawfishes which are similar in size to the red and often caught with it. The red crawfish has a thin and usually visible stripe that runs lengthwise of the tail along the ventral midline. This stripe, which is actually the ventral nerve cord, is dark blue. The similar slenderclaws do not have the pigmented nerve cord. Another feature useful in Texas for separating the Red from slender-claws is the presence of mostly light (cream or red) tubercles on the topside of the palm (not the ones along the inner edge). These are easy to contrast with the mostly black tubercles in similar position of slender claws.

Geographic distribution of the red crawfish, *Procambarus clarkii*. This species has been widely introduced out of its natural range into many areas in North America and elsewhere.

Habitat

The red crawfish lives in a diversity of habitats including streams, ditches, standing backwater, swamps and ponds. It primarily lives in surface waters but, in dry periods, it will burrow.

Distribution

The natural range of the Red is from northern Mexico to the tip of the Florida panhandle and north to southern Illinois and Ohio. Transfers via usage in aquaculture, bait fishing and hobby have resulted in introductions outside of the natural range.

In Texas, the natural range is throughout the eastern portion, along the coastal plain and upstream within rivers that flow from the Edwards Plateau into the Gulf. The red crawfish was originally described and given its scientific name based on specimens collected from a location west of San Antonio.

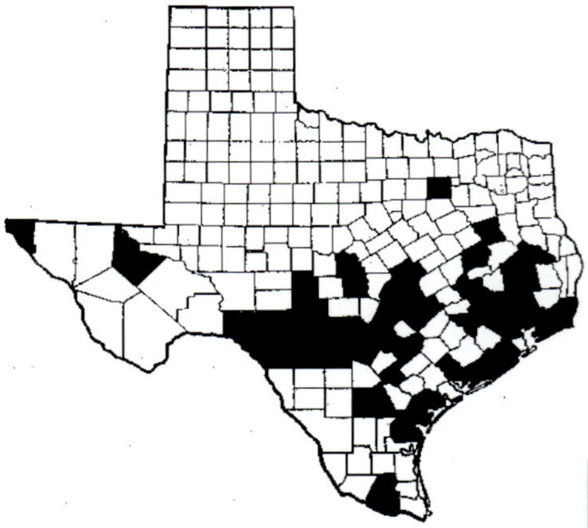

Texas counties with records of the red crawfish.

Catching the red crawfish

The red crawfish is caught in the smallest creek to the largest stream or lake. Like slenderclaws, they are frequently caught on a crawfishing trip with dip net, seine or trap.

Male *Procambarus clarkii,* Val Verde County.

Examples of *Procambarus clarkii*

Female, Val Verde County.

Female, Matagorda County.

Male, Kerr County.

Examples of *Procambarus clarkii*

Male, Kerr County.

Female, Jackson County

Juvenile male, Val Verde County

Procambarus - Southern Streamers

These crawfishes have handsome dark markings on the head and tail. They live almost exclusively in flowing waters. The group contains fourteen species, two of which occur in Texas. Adults are moderately large, reaching a size of more than 3 inches.

Recognition

The appearance of the two southern streamers that live in Texas, and many of those to the east, is almost identical. An examination of the gonopod is necessary for sure species identification. The following combination of features may be used to differentiate southern streamers from other Texas crawfishes. The rostrum is long and pointed and the areola is widely spaced. The black marking that saddles the hind end of the carapace extends forward along its sides. Two spines may be found on each side of the carapace at the backward edge of its cephalic portion and just below the black marking. The claws are very distinctive. The fingers have a black band midway of their length. In front of the band, fingers are cream colored to their tips. Prominent black tubercles on the outer, top surface of the palms contrast sharply with the tan background.

Habitat

Streams

Distribution

The range extends from eastern Texas to northern Florida, Georgia and South Carolina.

Geographic distribution of southern steamers.

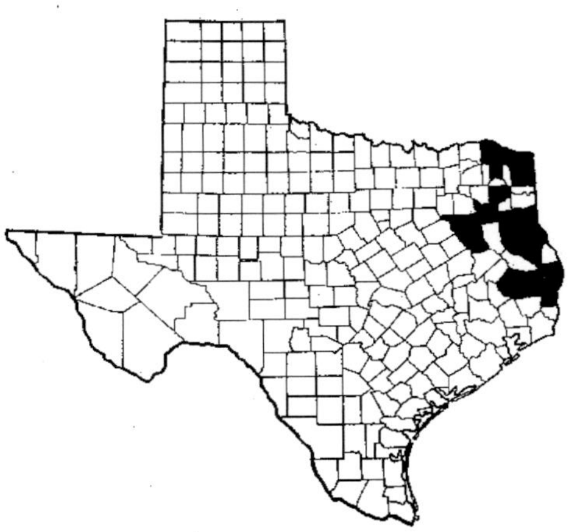

Texas counties with records of southern streamers.

Texas southern streamers

Texas has two species:

Procambarus
natchitochae
dupratzi

Common names used for Texas southern streamers: There are no established common names for these species.

Distinguishing Texas species

The general appearance of these two species is very similar. Examination of the gonopod structure is used to make the correct identification.

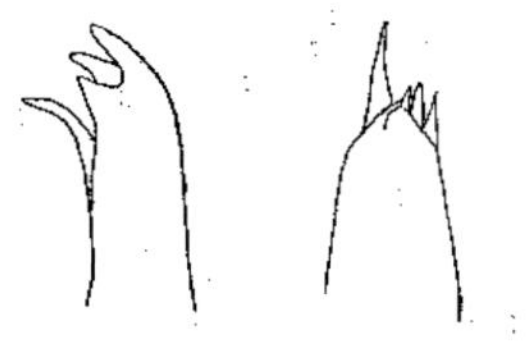

Form I male gonopod tips of Texas southern streamers (lateral view).

Distribution of Texas species

dupratzi	This streamer has the broadest distribution in Texas, found commonly in the Trinity, Neches and Sabine River Basins. It is also present in the Red River Basin.
natchitochae	The Red River Basin in the northeast corner of the state

Catching southern streamers in Texas

As their name indicates, these crawfish are caught in permanent flowing streams. Many of the small creeks within their ranges of northeast and eastern Texas have permanent flow. Shallow creeks are a good place to find and catch the streamers. A dip net and seine are good tools.

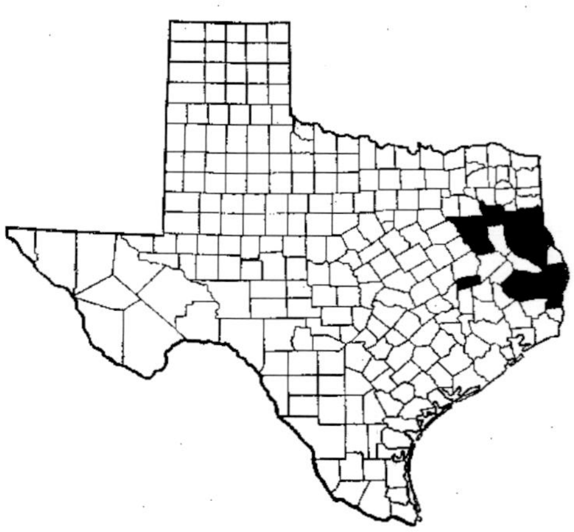

Texas counties with records of *Procambarus dupratzi.*

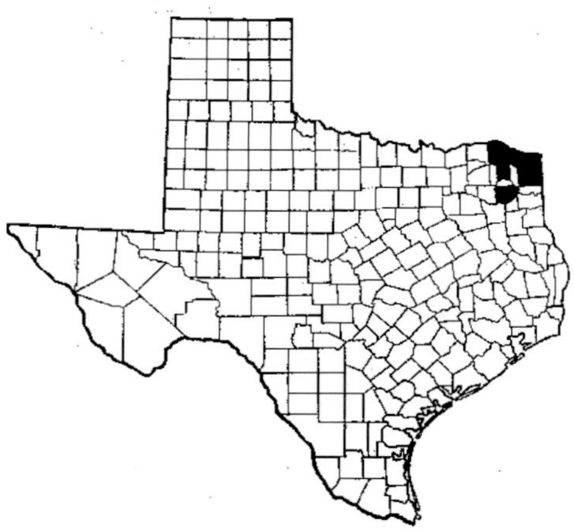

Texas counties with records of *Procambarus natchitochae.*

Male *natchitochae*, Bowie County.

Juvenile *dupratzi*, Polk County.

Male *dupratzi*, Smith County.

Male *dupratzi*, Smith County.

Female *dupratzi*, Smith County.

Procambarus - Grays

These species are the most abundant and widespread type of crawdad to be found in ditches of Texas. In most of the south-central United States west of the Mississippi River, people who catch crawfish for bait consider them to be two types. One type they call grays. These are commonly caught along with the other type which they call reds (the red and slenderclaw crawfishes). Medium sized grays are grayish to brownish in color (2 to 3 inches long), whereas "reds" of similar size have already begun to develop reddish coloration. Adults are commonly 2.5 to 4.3 inches in length.

Geographic distribution of gray crawfishes.

Recognition

Male Texas grays have hooks on the third leg pair (claws counted as first leg pair), instead of the third and fourth as is the case for most other Texas *Procambarus*. Only the blackgirdle hairclaw, the Navasota crawfish, the creekbank crawfish and the prairie burrowers share this feature with them. The lack of a broad, dark stripe located top-central on the tail is useful in separating grays from other large crawfish such as the slenderclaws and the red crawfish. Depending on the species, older, large grays have a body coloration that is some degree of red, green, blue, or lavender. Such coloration is most apparent on the claws. Older animals also have more robust claws.

Habitat

Grays are usually found in periodically disappearing surface waters. When water recedes, they use burrows. Ditches, backwater pools and wet-weather streams are examples. In central, eastern and coastal regions of Texas, grays populate low areas and ditches that are a long distance from a permanent stream. Chimneys, usually 6 inches or less, are common. In west-central Texas where the water table is more closely associated with permanent water, they are more often found near, or within, permanent streams and ponds.

Distribution

Grays cover a broad range in central North America and occur almost everywhere in Texas that is suitable for crawfish life.

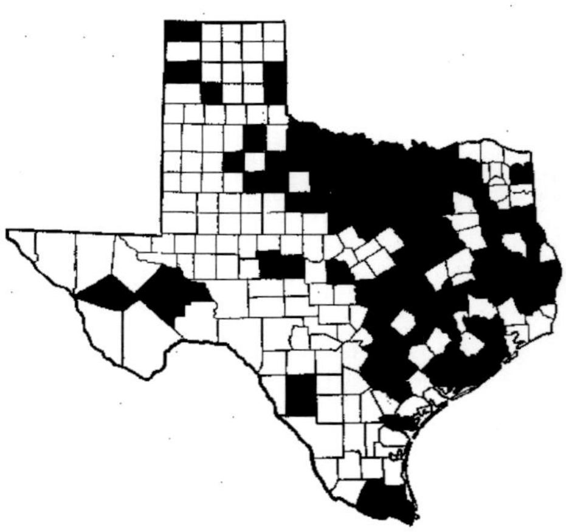

Texas counties with records of gray crawfishes.

Texas Grays

Texas has three grays:

Procambarus
- *curdi*
- *simulans*
- *regiomontanus*

Common names for Texas grays: The species have no common names.

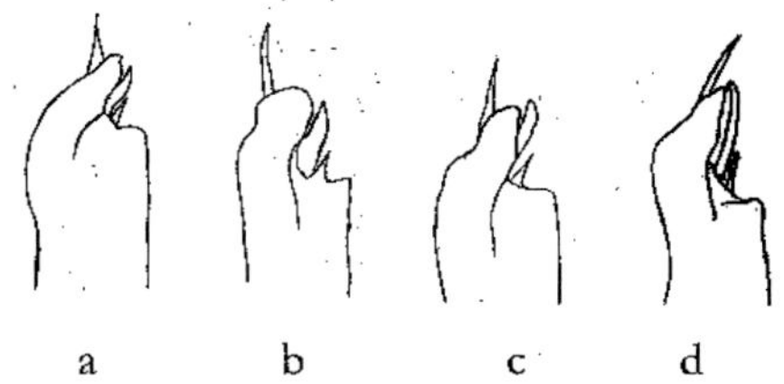

Form I male gonopod tips of Texas grays (lateral view). a. *simulans*. b. *curdi*. c. *simulans*, the form found in western and northwestern Texas. d. *regiomontanus*.

Distinguishing Texas Grays

Young and old *curdi* and *simulans* have almost the same appearance. Examination of the gonopod is the only practical way to distinguish them. A lateral view examination of the gonopod of *curdi* reveals that the caudal process (the broader terminal element shown in the figure) is positioned so that its flat side faces the observer. When viewing the caudal process of *simulans* in the same manner, the flat side appears not a broad and not facing the viewer

Some crawfish specimens that look like *curdi* and *simulans* occur west of the Gulf Coastal Plains. We include them as *simulans,* even though the gonopod varies somewhat from the average *simulans* specimen. The central projection (the finger-like projection to the right of the caudal process in the figure) extends beyond the caudal process.

A gray crawfish that lives in the Rio Grande Valley is similar enough to *Procambarus regiomontanus* to include it in that species. The established range of *regiomontanus* is northwestern Mexico. The claws are usually brick red in older females and bluish to red in males. Mature males have a rostrum that, for grays, is relatively long and narrow.

Distribution and habitat

curdi — *Procambarus curdi* occurs in eastern Texas, eastern Oklahoma and southwest Arkansas. In Texas, we have caught this species as far west as Brazos and Collin counties. Habitat preferences seem to be ditches and low areas, often far from permanent surface water. They spend most of their time underground when temporary water disappears. In daytime, juveniles are found in temporary water. Adults emerge mostly at night.

simulans — *Procambarus simulans* has the broadest distribution of the grays. Its range covers much of the lower central U. S. Anywhere that *simulans* can easily reach the water table is a likely place to find one of its land burrows. In the drier areas of western Texas and Oklahoma favored burrow sites are within, or adjacent to, permanent water.

regiomontanus — This species occurs within, or nearby, the lower Rio Grande (Rio Bravo) basin and is common in ditches and irrigation canals. Like *curdi* and *simulans*, it is easily caught from ditches with water.

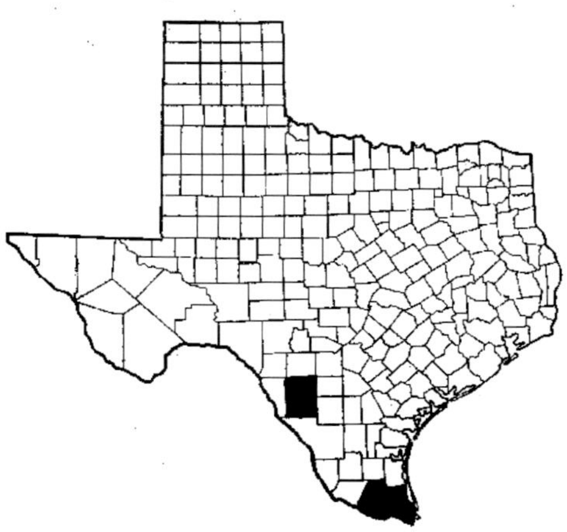

Texas counties with records of *Procambarus regiomontanus.*

Catching Texas grays

A dip net is an excellent tool for catching grays. A good place to find these crawfishes is in standing water of a shallow ditch. Such ditches occur where the ground is level. Ideal ditches dry up slowly and are located more than half of a mile from a permanent stream. Using a dip net in muddy pools shallower than 2 feet during the daylight hours often results in medium sized, and occasionally larger, catches. When dipping in the muddy water of a ditch, try to work the dip net quickly. Being a burrower first, and surface water dweller second, a gray makes a fast retreat to its underwater tunnels once the water is disturbed.

The best time to catch large grays in central, southern and eastern Texas is during a night that follows a rainy period. Have knowledge of their location from dry period checks of roadside drainage ditches. Once a good burrow location is known, return to it after a rain. Grays can usually be seen nearby their burrows in standing, clear rainwater only a few inches deep. If pools flood to over than 1 foot, a trap is usually an efficient catching method.

Male *Procambarus simulans*, Brazos County.

Male, Collin County.

Male, Brazos County.

Juvenile female, Brazos County.

Female, Freestone County.

Female, Collin County.

Male, Brazos County.

Female, Brazos County.

Male, Lee County.

Male, Collin County.

Male, Somervell County.

Male, Brazos County.

Female, Collin County.

Male (western-northwestern form), Collingsworth County.

Male, Willacy County.

Juveniles, Hidalgo County.

Female, Hidalgo County.

Procambarus - Creekbank Crawfish

The creekbank crawfish, *Procambarus kensleyi*, is found in eastern Texas, western Louisiana and southern Arkansas. In Texas, it is found in the Neches, Sabine and Cypress basins. This crawdad burrows into the banks of permanently flowing creeks. Spots are apparent on the surface of the carapace where it covers the gills. The areola is broad, a feature common to crawfishes that inhabit flowing waters. Adults grow to a relatively large size of more than 3 inches. Holding this crawfish is somewhat difficult because it can extend the claws far backward. We propose the name creekbank crawfish for this species, indicating a habitat preference mentioned by early investigators.

Distinguishing the creekbank crawfish

The body coloration of *Procambarus kensleyi* is brown to gray. The broad areola is very distinctive. Within its Texas range, the only other *Procambarus* with a similar appearance and a broad areola is the blackgirdle hairclaw. These may be distinguished by the structure of the rostrum. Texas grays are very similar to *kensleyi*, but they have a narrow areola.

Habitat

Procambarus kensleyi lives in creeks and burrows of creek banks. A shallow pool of a small, flowing creek is a good place to catch this crawfish. Traps work well in deeper, nearby ditches.

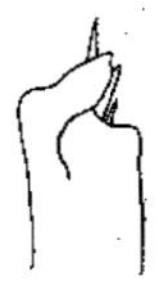

Form I male gonopod tip of *Procambarus kensleyi* (lateral view).

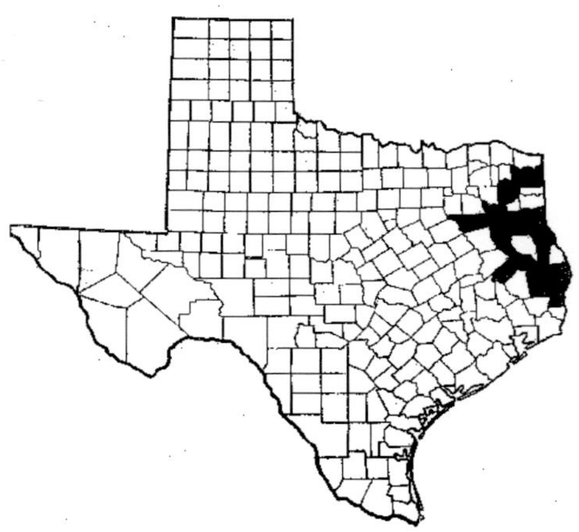

Texas counties with records of *Procambarus kensleyi.*

Male, Smith County. Note spots visible on carapace side.

Male, Camp County.

Female, Camp County

Procambarus - Hairclaws

Hairclaw crawfishes have a thick row of fine hairs (setae) on the inner side of the palm. This row of hairs is called a beard.

Recognition

Mature males and females have a beard on the inside margin of the claw.

Habitat

Hairclaws live in burrows and nearby, standing, or running waters.

Geographic distribution of hairclaw crawfishes.

Distribution

Hairclaws occur in two geographical ranges. One range includes Texas, Louisiana and southern Arkansas. The other includes the coastal plain from South Carolina to northern Florida and eastward to eastern Louisiana. Two species occur in the Texas, Louisiana and southern Arkansas range: *Procambarus tulanei,* (LA and AR) and *Procambarus nigrocinctus,* (TX). Several burrowing species in eastern Mississippi have tufts of setae on the inner palm, but the setae are not formed into a full beard. Species of another group that lives in southern Mexico, Belize and Guatemala have hairy fingers or palms that are not organized into a true beard.

Texas hairclaws

Texas has one hairclaw:

Procambarus
nigrocinctus

Common name: No common name is used for the species. *Nigrocinctus* is Latin for "black girdle."

Recognition

Procambarus nigrocinctus has a dark band that colors the entire back edge of the carapace. The beard develops in mature animals, more so on males than females. Juveniles have little or no beard, but the black band is obvious. *Procambarus curdi* and *Procambarus kensleyi* have similar appearance to *nigrocinctus*, and they live within its range. The broad areola of *nigrocinctus* is useful in separating beardless *nigrocinctus* juveniles from *curdi.* Though it has a broad areola like *kensleyi*, the rostral tip is much more developed in *nigrocinctus.* Juveniles also show the distinctive black girdle pattern. The speckled body and red fingertips of *nigrocinctus* adults give them unique beauty.

Form I male gonopod tip of *Procambarus nigrocinctus* (lateral view).

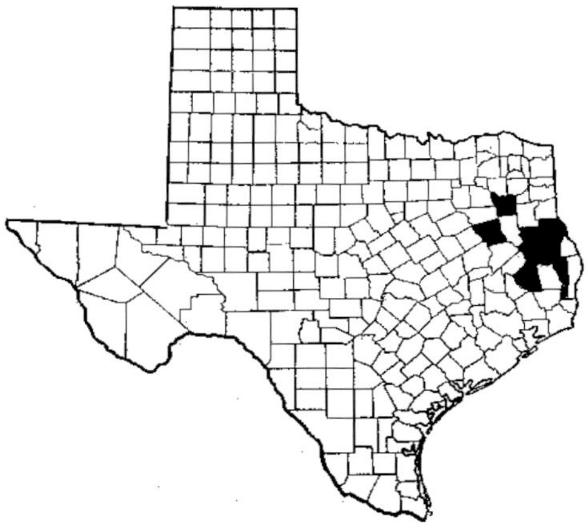

Texas counties with records of *Procambarus nigrocinctus.*

Habitat

Although *nigrocinctus* is usually caught in flowing streams, we have seen shed exoskeletons in the bottom of dry, temporary pools some distance from a stream. In Louisiana, *tulanei* is known to burrow near streams and into stream banks,and has been captured in flowing water.

Distribution

The blackgirdle hairclaw lives in the Neches River basin of eastern Texas.

Catching the blackgirdle hairclaw

Most reported catches of the blackgirdle hairclaw are from flowing creek water. We used dip nets to catch it from flowing water. A seine or dip net could probably be used to catch it from the overflow water of a floodplain.

Male, Angelina County.

Female, Angelina County.

Male, Angelina County.

Procambarus - Prairie burrowers

Prairie burrowers are handsome crawfishes of medium size. They inhabit lands which were once prairies. There, they live in, or near their burrows. Adult length is 2 to 3 inches.

Recognition

The lower portion of the carapace and lower margin of the tail, back to the tail fan, is seen in side view to have cream coloration. The upper portions of the carapace and tail have a background color of chocolate brown in females, and tan-green to red in males. In both sexes, a light stripe runs along the top part of the tail and onto the telson. Claws are brownish to purplish-blue in females and tan to blue in males. The general appearance of the different species is very similar. Species assignment is based primarily on gonopod structure.

Habitat

The burrows are usually found near or within tree stands of prairie lands, but in some places, burrows completely cover fields. Burrow areas may be far from places where rain would regularly produce pools of temporary water. Tunnel depths may exceed 6 feet. Such depths are perhaps necessary for survival, because in severe drought, the area's soil may dry to nearly 10 feet. Excavation deposited around a hole is more often seen as a plume or low hill than as a chimney. Foraging occurs in the vicinity of the burrow at night, especially during wet or warm, humid weather. Breeding behavior is also thought to occur in similar conditions.

Distribution

Prairie burrowers occur from northeast Texas and southwestern Arkansas northward into the plains, around the Ozark-Ouachita Highlands and into Iowa, Illinois and southeast Wisconsin.

Geographic distribution of the prairie burrowers.

Texas prairie burrowers

Procambarus
steigmani
regalis

Common names for Texas prairie burrowers: No common names are used for these crawfishes.

Distinguishing Texas species

The appearance of both species is basically the same. The gonopod should be examined and compared the gonopods of specimens with verified identifications. According to Hobbs (1991), the gonopods of adult male *steigmani* are not parallel like those of *regalis*. They are positioned with one gonopod not reaching as far forward as the other (not symmetrical).

Distribution and habitat of Texas species

Collection localities of these upland crawfishes are too few in number to indicate much about the distribution of Texas species. The known Texas distribution is near to Oklahoma. *Procambarus steigmani* is known only from Collin County. *Procambarus regalis* is previously known from southwestern Arkansas. The habitat is like that described for other prairie burrowers.

Catching Texas prairie burrowers

Being nearly open field burrowers, prairie crawfishes are difficult to catch. Pit traps may be used. They do come to the surface and look out from their burrows, particularly on wet nights, but they are very quick to retreat if startled. Very small juveniles are occasionally collected from a ditch near a burrow colony.

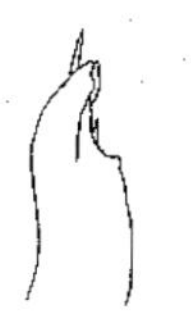

Form I gonopod tip of *Procambarus regalis* (lateral view). The structure of the gonopod tip of *Procambarus steigmani* is essentially identical to that of *Procambarus regalis.*

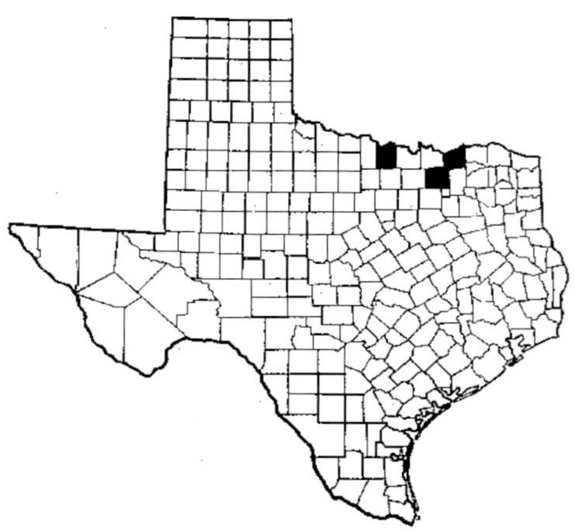

Texas counties with records of prairie burrowers.

Male *steigmani*, Collin County.

Female *steigmani*, Collin County.

Female *steigmani*, Collin County.

Male *regalis*, Fannin County.

Procambarus - Navasota Crawfish

This medium size crawfish is a primary burrower. Its known range is a small area near the Navasota River in central Texas. It is rarely caught. The crawfish was named *Procambarus caeruleus* by Drs. J. Fitzpatrick and M. Wicksten. The "*caeruleus*" indicates the color of the sky blue claws. Sky blue claws are not very apparent in the specimens we caught, but the legs and claws do have a hint of blue or blue-green. Dr. Wicksten (personal communication 1/28/05) stated that bright blue claws are found only occasionally on older breeding males. We propose the common name, Navasota crawfish, for this unique animal.

Recognition

The broad claws of older animals are similar to those of fatclaws. A series of oblique markings extends down each side of the tail. The body is tan brown. The gonopod structure is very unique but most similar to that of grays.

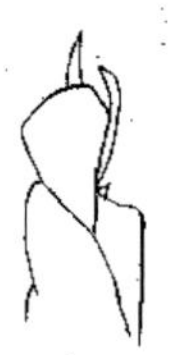

Form I male gonopod tip of *Procambarus caeruleus*, the Navasota crawfish (lateral view).

Distribution and habitat

The Navasota crawfish is found to the west of the Navasota River in eastern Brazos County, Texas. A typical burrow entrance has only a little excavated soil around the opening and remains plugged except during very wet weather.

Catching the Navasota crawfish

We have caught this crawfish with dip net during the night from shallow pools of the Navasota River flood plain. We also caught a number of specimens using a dip net during daytime and after heavy rains from ditches.

The Navasota crawfish, *Procambarus caeruleus*, occurs in the Navasota River drainage of the county indicated, Brazos County, Texas.

Female *Procambarus caeruleus*, Brazos County.

Male, Brazos County.

Juveniles, Brazos County.

Male, Brazos County.

Male, Brazos County.

Male, Brazos County.

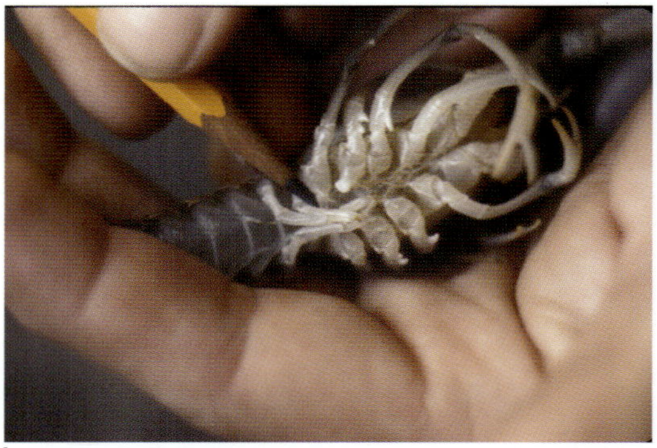

Examples of variety of appearance in crawfish. Top row, left to right:: A *Procambarus simulans* specimen that has darkened during the intermolt phase. The same animal, after it has molted. A freshly molted *Cambarus diogene*s specimen showing a bright coloration and a left re-growth claw. Second row, left to right: A *Procambarus simulans* specimen that has undergone a late-life conversion to a partial blue coloration. *Faxonella clypeatus* females showing striped and unstriped color morphs. Third row, left to right: A specimen of *Orconectes deanae*. that is a blue color variant. One gonopod of a Form I male (and the two modified swimmerets immediately behind it) being lifted with a pencil. The dark tip indicates that it is a Form I male. Bottom row: A microscopic image of the distal third of a gonopod from a Form I slenderclaw crawfish. Note that the brownish (amber) coloration of the tips is indicative of a Form I male instead of a Form II male (without well-developed terminal elements of brownish color). In the order described above, the percent of actual size of the images is 70%, 70%, 85%, 45%, 50%, 77% and 60%. No scale is given for the magnified gonopod.

Example of variety of appearance in crawfish. Above: A *Procambrus simulans* female in the intermolt stage three days before molting. Below: The same crawfish as it appears soon after molting. Note lighter coloration and increase in size of claws.

A comparison of the sub-groups of *Procambarus* using body features of adult males

Hooks present on legs of third and fourth pair (claws counted as first pair)

Longpalms

Longpalms are small, reaching only about 2 inches. Two dark stripes run along the back from eye to tail. The areola is broad. The palm is elongated, becoming more so with age.

Slenderclaws

Slenderclaws are large crawfish, becoming adults at around 3 inches and growing to 5 inches (usually 4+) long. The palms are narrow and fingers are long. A broad, dark stripe is usually present on the top of the tail. Some specimens show a dark stripe on the side of the carapace which, if very obvious, may continue along the side of the tail. The areola is narrow.

Red crawfish
The red is another large crawfish, attaining Form I stage by about 2.5 inches and reaching more than 5 inches (usually 4+) long. A broad, dark tail stripe is apparent in most mature individuals. The claw is not as slender as a slenderclaw male's and, instead of having black tubercles arising from the back of the palm, it has most, or all, tubercles with light coloration. The areola is closed or very narrow.

Southern streamers
These are moderately large crawfish. Breeding males are found at 2.5 inches total length and the largest individuals growing to more than 3.5 inches. Dark striping is obvious along the sides of the tail and carapace. At the rear end of the carapace the bold stripe continues upward, narrowing toward the top. These markings and the tubercles on the claws are black and contrast strongly with a relatively pale body background. The fingers are also dark with cream tips. Two spines are present on each side of the carapace at the junction of head and thorax. The areola is broad.

Hooks present on legs of pair three (claws counted as first pair)

Grays
Grays are large crawfish. The Form I stage is reached at about 2.5 inches. Growth is to more than 4 inches. Stripes along the body surface are usually faint and thus not a dependable feature for assignment of specimens to the group. Tubercles on the back (top) of the palm are dark. The areola is narrow.

Creekbank crawfish
The creekbank crawfish grows to more than 3 inches and has a broad areola. The shell coloration is brownish-gray. The portion of the carapace that covers the gills is usually covered with spots.

Hairclaw
The blackgirdle hairclaw is essentially a gray with a beard on the inner margin of the palm. The beard develops with age becoming fully developed when the crawfish attains a total length of about 3 inches. The body surface has a mottled or speckled pattern and a dark "girdle" at the rear margin of the carapace. The areola is broad.

Prairie crawfishes
Prairie crawfishes are medium sized. Adult length is 2 to 3 inches. A dark coloration on the topside of the carapace gives way to a creamy color midway down the carapace side. Texas species have a showy blue, purple, red and brown coloration. Claws are more robust, like the fatclaw crawfishes. The areola is narrow.

Navasota crawfish.
This crawfish is medium sized. Adult size is 2 to 3 inches. The body coloration lacks obvious patterns. The claws are very broad and the fingers relatively are short. Claws are brownish to pale blue-green. The areola is narrow.

Crawdad Distribution

The presence of crawfish means that they are able to settle and survive in a place. The place must satisfy their environmental needs, meet food requirements and provide suitable protection. The essential requirement is that freshwater be available either above, or below ground. Humid conditions will do for some kinds, at least for a while, if the gills do not dry out.

Nearby to permanent surface waters and within low areas where water is captured and settles without quickly draining away are good places for crawfish to build their miniature "dug wells." Land features and water barriers that occur below, or at the earth surface sometime force water nearer than usual to the surface. Such places have shallow water tables, seepages or spring flows favorable to burrowers.

Changes in the general landscape and pumping from wells over the last century have caused a lowering of the water table and drying up of many springs in Texas. Some subterranean crawfish species unknown to man could have become extinct. Changes brought by human progress may have caused other species to fare better than in earlier times. For example, abundant habitats were provided by construction of earthen irrigation ditches and drainage ditches along roads and railways.

Distributions are determined by the spread and continual reestablishment of crawfish species within areas of suitable conditions. Dry years and wet years are common in the state. Some regions are drier than others in particular years. On average, evaporation exceeds rainfall westward of central Texas. The dry climate of western areas does not favor a general presence of crawfish. There they occur only in, or in close proximity to, permanent streams and large impoundments. In contrast, across central and eastern Texas, many lowland areas with a shallow water table exist. There, crawfish are often found great distances from permanent waters in temporarily flooded ditches and other low spots. The prairie crawfishes are an exception, settling far from permanent water in higher terrain where they dig deep into the earth.

Some crawfishes are known to migrate overland and up streams. This is a notable behavior of the red crawfish, the slenderclaws and the grays. We have seen numerous red crawfish moving up streams in west Texas. When a dam is met, they move onto land and walk parallel to the stream, beginning perhaps a quarter mile from below and often not within sight of the water. Red crawfish and gray crawfish are occasionally seen wandering on land after rainfall in no particular direction. Fatclaws wander from their burrows during humid nights.

Natural events have the potential to aid reestablishment and spread of crawfish to distant areas. New places may be reached when flood waters cover large areas of a lowland or flood plain.

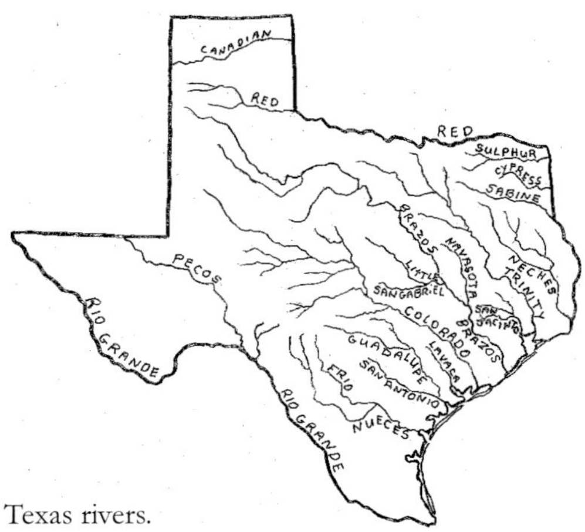

Texas rivers.

Geographical distribution

We now present a simplified geography of various environmental regions and resident crawfish groups. [Many maps and classifications have been constructed to portray the environmental diversity of Texas. A widely available reference that clearly presents Texas geography is the Texas Almanac, an annual production of The Dallas Morning News].

A coastal flatland skirts Texas along its contact with the Gulf of Mexico. It extends two or three counties inward and runs northeastward from Corpus Christi to the Louisiana state line. Much of the area is a prairie with patches of live oak, except near small streams and rivers which support small to large expanses of woods. As this region extends northeastward into rice country, the presence of irrigation canals for agriculture increases. Frequent rainfall near the Gulf supplies an abundance of water to ditches and wetland areas. Gray, slenderclaw, red, pygmy, fatclaw and longpalm crawfishes are abundant in this region. The small portion of flatland present around the outflow of the Rio Grande could be considered an adjunct portion of this same region. Gray, red and slenderclaw crawfishes are present there.

The Gulf Coastal Plain includes and extends above the coastal flats. This large area of hilly, rolling terrain extends from the Rio Grande into eastern Texas. From there it continues to the Atlantic Ocean. Within Texas, the most eastward portion, north of the coastal flat and south of the Red River, is the westward limit of the southern pine forest, commonly known as the Piney Woods. The area's relatively high rainfall supplies an abundance of permanent small creeks, wetlands and artificial impoundments and supports a shallow water table in places where level terrain exists. Rivers and larger tributaries have overflow areas that accommodate hardwood bottoms. Small creeks that cease to flow during dry times usually retain water in pools under forest canopy and at culvert outflows. In the Piney Woods' southern portion, longpalms and pygmies are common. Southern streamers, the hairclaw crawfish, the creekbank crawfish and grippers live in the central and northern portions. Slenderclaws, fatclaws and the red crawfish are present throughout. Orcos are rare.

In southwest-central Texas, the Gulf Coastal Plain meets a plateau that rises abruptly along a fault line. Above the fault lies what is known as the Texas Hill Country, an area of hilly relief and unique vegetation. The cypress-lined rivers that flow through the area are supplied mostly with spring water. The water flows clear over a rock-cluttered limestone base. Tributaries to the Colorado and Brazos Rivers descend from the eastern portion of the plateau. The upper portions of the Guadalupe River and the upper portions of the Nueces River flow southward through the plateau. The crawfish population of the streams of the western Hill Country consists of the red crawfish and orco species. Gray crawfishes move up onto the plateau along the edges of the more eastward portion.

Returning to the Gulf Coastal Plain, the portion northeastward of the Rio Grande and southward of San Antonio is a brushy or chaparral area. Gray and red crawfish are present near and along the Rio Grande and slenderclaws are present near to the Gulf.

Texas land regions.

Northeastward of the chaparral area to the pine forest of eastern Texas is a large area of rolling plains and prairies [human modification has greatly changed the natural or original condition of Texas prairies]. Much of the area features blackland soil that is worked for crops and pasture. This blackland prairie stretches north to south and extends westward around the base of the Hill Country to San Antonio. Prairie portions topped with lighter soils are mostly used for pasture. These are usually infested with some degree of mesquite overgrowth.

When advancing from the chaparral to the northeast, one encounters an increase in patches of woody vegetation consisting largely of mesquite, elm, live oak, sugarberry and post oak until a more confluent post oak forest (post oak, blackjack oak, water oak, sugarberry (hackberry), elm, cedar and others) is met. Streams coursing southward through the prairie to the coast are lined with woody vegetation (willow, pecan, elm, sugarberry and cottonwood). Some larger streams of the southern portion of the post oak forest have flood plains supporting the growth of bottomland hardwoods. Within the southern portion of the post oak forest area, slenderclaws, the red crawfish, grays, pygmies, longpalms, the Navasota crawfish and fatclaws may be found. Orcos are rare. From the southern portion of the post oak forest to the Hill Country slenderclaws, red crawfish and grays predominate. Within the northern portion of the post oak forest area, slenderclaws and grays are the principal crawfishes.

From the northern post oak area across to Texas' other abrupt rise in elevation (the High Plains) is land of flat and rolling prairies, strips of forest and occasional places of substantial relief. Gray crawfish live here. The land becomes gradually drier and brushier toward the west. Most streambeds go dry in summer. Grays, however, are common in ditches that contain temporary water.

An abundance of livestock watering ponds occurs throughout central Texas, south of Dallas and into the post oak forest. Around and in the ponds are grays, slenderclaws and the red crawfish.

Still farther north and eastward near to and within the Red River basin, crawfish diversity increases. True prairie crawfishes are known to occur northeast and northwest of Dallas-Forth Worth. Grays and slenderclaws are common in the area. The devil fatclaw is rare. Orcos, grays and grippers are found within the Red River and its tributaries.

The high plains rise from the rolling plains in northwest Texas. Grays and orcos dwell there in the permanent water of rivers and reservoirs. The Edwards Plateau, its most eastern portion considered the Hill Country, meets the high plains on its northwest portion. Orcos and the red crawfish populate the plateau's permanent streams and reservoirs. Farther west, from the Edwards Plateau to El Paso, is an arid, mountainous region. There, grays and red crawfishes are found in permanent water.

Catching Crawdads from Open Waters

Crawfish are normally captured either from surface waters or burrows. Only occasionally are they picked up when crawling across the ground or climbing up stream banks.

Open (surface) waters include ditches, sloughs, ponds, streams, lakes and flooded lowlands. Catching crawfish in surface waters is most enjoyable and productive in a water shallower than 1 foot. This depth is suitable for knee boots. A depth of 1.5 foot is getting too deep and reaching to the brim of an adult knee boot.

Dip Net

A long-handled dip net with a strong D-shaped ring is the equipment of choice. A good total length for net and handle is 4.5 feet. A good handle diameter is 1.1 inch. If a long-handled net is not available, one may rig a net by attaching a short pole piece to a short handled net (i.e., 2 .5 feet total length) with a piece of PVC pipe that is used as a sleeve. A suitable mesh size for retaining the smallest varieties of crawfish is 0.25 inch. A dip net may be worked from shore or while wading.

When no water exchange is present, shallow surface water often lacks clarity and the pool bottom cannot be seen. The added darkness given to the substratum by the turbid water during daytime provides protection to crawfish and may enhance their activity. Crawfish living among bottom litter and vegetation are easily caught with a dip net.

Always plan to use surprise when approaching standing water such as that in a ditch. After selecting a good spot, plunge the dip net to the bottom, just beyond the chosen place, and quickly drag it across the bottom toward the bank where you stand. Small areas, for example, a ditch pool of 5 feet, can sometime be thoroughly dipped in seconds by using multiple dips in rapid succession from end to end. Otherwise, a series of drags is made while wading, lifting the net to view its contents after each dip or, if litter or weeds do not build up in the net, after a series of rapid dips. Dip nets can also be used to pull along a weedy shoreline or immediately dipped in a place where solid cover is removed. A mass of plants and dead matter from the bottom usually is present in the net bag. These can be removed without much risk of losing captured crawfish, because they are usually found at the bottom of the bag.

If the water is clear and the crawfishes are easily seen lying on the bottom, a dip net can be quickly drawn from behind them, along the bottom and while making adjustments for their escape maneuvers. In clear flowing waters, a dip net may be used in conjunction with a dark stick. Movement and touch of a dark-colored stick does not easily stimulate a rapid escape. The maneuver of such a stick may sometimes be used to push the animal into the net.

Hand Grab and Dip Net

Shallow, flowing waters are often clear and allow catching of crawfish by hand or in conjunction with a dip net. Upon arrival at such places, it is a good idea to take a short time to spot any crawfish on the bottom before disturbing the water. A quick dip will often catch a good one by surprise. If a crawfish sees you and tries to hide, it may be hand pinned to the bottom before it swims away. To do so, ease up to it and thrust the hand quickly down on top with a force that is fast enough to work, but gentle enough not to crush the creature. Hand catching and dip netting work well together in rocky streams. Crawfish usually hide under limbs, logs or sticks during daytime. When such cover is lifted, one may immediately swim out and away, usually toward deeper water. The dip net may be used to catch it at that moment. Often, until the flow causes the water to clear, the crawfish remains in place until it can see what is going on. Usually, you can see it before it sees you, or decides what to do. A hand grab is sometimes best because the net dip may miss due to uneven bottom and obstructions. If the water is clear and the current is strong, disturbing rock cover results in crawfish being swept down stream in the current as they abandon the cover. Position the dip net immediately below the selected rock before moving it. It is best to wade against the flow in streams with silty bottoms so that the catching area remains clear. If rocks are over-turned, remember to turn them back over onto the original place. This preserves the natural look of a clear stream.

Seine

A 10 X 4 feet minnow seine of 3/8 inch mesh is a useful device to drag through places that are mostly free of roots, rocks and thick vegetation. A seine may also give good results when it is set in place in riffle areas to catch crawfish as they leave cover and move downstream in swift flow.

To rig a seine, attach corner ties to two pull poles. The poles should be about 5.5 feet each in length and have a 1.5 inch thickness. Cut notches at the attachment sites to prevent slipping. Attach ties of the weighted line to ends of the poles. Connect the "float" line ties a short distance from the other pole ends so that a slight sag forms toward the seine midpoint when it is held up like a blanket.

Two people normally use the seine while spread apart at a distance that causes the top and bottom seine lines to slightly bow and the netting between to form into a porous basket. A pole is held with one hand at mid-length and the other at a point between the float line connection and the upper end. The seiners push or pull the seine by holding the poles at an angle with the bottom. The weighted or "lead" line is kept taught as it and the lower pole ends move across the bottom. The weighted line thus moves somewhat ahead of the float line.

Loss of content over the seine top while seining is avoided by keeping the float line taught enough and forward enough to maintain the basket form. An effort is also made to avoid loss of content under the lead line (a "roll under"). This is done by keeping the line taught, holding the poles at an angle of less than 75 degrees to the bottom. If the lead line encounters an insurmountable snag, it is best to immediately lift the net in blanket-spread fashion by bringing the poles parallel to the bottom while lifting and pulling away from each other. This can be similarly done when a large content of debris accumulates in the net. If an obstruction is noticed in advance of the seine, or if the clarity of the water allows seiners to see good crawfish or interesting materials enter the basket, the seiners agree to lift the seine simultaneously and make the catch.

Usually, the seine is pulled successfully to shore, either directly or indirectly, by one person pivoting at a spot while the other continues to swing around. When seining into shore, attention is given to land the lead line in a way that does not allow loss of content under it. Again, this is done by keeping the lead line taught and watching for snags. Often, when landing the seine, shoreline vegetation stops the seine a small distance from the water line. If this distance is short, a visual inspection of the shallow water will detect presence of crawfish that may be taken by hand. If not, or if the water is too muddy, shored crawfish may be enticed to enter the seine basket by quickly using the feet to slosh edge content into the seine basket while still holding it in place.

A seine may also be used in flowing streams. The stream is quietly entered below the shallow stretch of interest and the seine is anchored by thrusting the poles into the bottom or by having it held by companions. Someone then kicks or disturbs the bottom immediately upstream from the seine to encourage the crawfish to make an escape. They are then swept downstream into the seine.

An individual may use a seine in a similar manner if it is made shorter by rolling the ends around the poles. The poles are held at mid length and at a proper angle while the seiner walks backward and upstream ahead of the seine.

Again, seines are best used in places with suitable open water space with smooth, snag-free and obstruction-free bottoms. A dip net is overall a better device for shallow water crawfishing.

Traps

Traps are best for catching crawfish in deeper water, though they are also effective in shallows. In some cases, they may be the only good way to catch crawfish in vegetation-choked water. Various kinds of traps are available, most with some type of funnel opening to the inside. The most widely available and inexpensive kind is a minnow trap with a cylindrical shape and inverted cones (funnels) on each end. The diameters of funnel openings determine the size of crawfish able to enter. Some state fishing regulations place limits on the opening diameter, so that only small fishes may be caught. Regulations usually do not place limits on the mesh width of trap walls but mesh size is important in trapping small crawfish species. Bait of all sorts is used: bacon, other meats, fish, pet food, etc. The bait should not easily disintegrate in water. The bait cannot be reached through the mesh if it is kept away from a trap's sides. For example, bacon or cheese wrapped in nylon stocking and suspended inside avoids such pitfalls and resists destruction by crawfish already caught. Cat food contained in small cans is a good choice and is convenient. The lid is partially opened to retain content and merely tossed into the trap. Traps are usually set out of view from passers-by. Besides their curiosity to check your trap for you, they may find it tempting to snitch the trap while you are away. Using a dark retrieval line between the trap and the bank attachment is a good idea.

Other Methods

A lift net may be useful for occasions when one intends to give close attention to catching crawfish. A lift net is constructed of a sheet (usually square) of netting or wire mesh attached to a frame at the margin. Lines affixed to corners of the frame attach to the end of a pole. One sets a lift net by allowing it to sink to the bottom with its bait attached top center. After a while, or when crawfish are seen in the net, the contraption is lifted up and out of the water with the pole and then set onto shore. A sag in the netting deters escape. Another design uses two sturdy "V" shaped wires affixed together at their apex and spread perpendicularly, so that the ends attach to the corners of a square piece of netting. Basket traps meant to catch coastal crabs may be used also for crawfish if the netting mesh is narrow enough to retain crawfish.

A baited line tied to a pole may be used to gently lift crawfish to the water's surface. Once at the surface, they are quickly dipped with a net. Chicken necks and bacon may be used with a weight affixed to the line near the bait. Catching crawfish this way is a delicate skill. They tend to grasp the bait with the claws and then pass the bait to the smaller walking legs and mouth. Crawfish and bait part ways rather quickly if one does not gently pull the crawfish to the surface. To avoid such releases, the line may be pulled slowly along the bottom with bait and crawfish in tow toward the shore. Once visible in the shallow water of the shore, the crawfish is quickly grabbed by hand or scooped with a net.

Small containers may be used in areas where little cover is afforded crawfish residents. Cans or wide mouth bottles are set in places where they can be easily retrieved and left for one, or several days. Upon return, a careful retrieval of a series of these containers may result in several good specimens.

Night Catching

Crawfishes that spend their lives in surface waters move about mostly during nighttime. Beginning at dusk, they move from under their daytime cover or emerge from underwater burrows to feed and interact with other crawfish. At night, a headlight may be used to find crawfish in shallow clear water. They may be caught using a dip net or by hand. The eyes shine like rubies, helping one to locate them. Hunting is easier if the illumination device produces a broad beam of light. Areas to be searched with lights should be visited beforehand during daytime to determine their suitability for crawfish catching. A pre-visit is also important to locate hazards and to mentally chart the easiest path to course in darkness.

Using a "D" rim dip net.

Got one.

Catching Crawdads from Burrows

During, or just following, rainy weather is the best time to catch burrowers, day or night. If rainfall provides enough water to cover the burrow openings, crawfish will emerge to enjoy the occasion.

One should seek out active burrows, ones that show signs of current use and remodeling activity. Fresh mud around the burrow opening indicates that the crawfish is home. During remodeling, they push wet mud to the surface, drop it off and return below for the next load. Excavations are done mostly at night but crawfish may sometime be found digging during overcast days.

Daytime methods

Catching burrowers from air-exposed burrows by day is challenging, but can be done. They, of course, can be removed by digging to the burrow bottoms but this takes much time and effort and is no fun. When active at the burrow opening during cloudy days, most crawfish are easily startled by visual and physical disturbances. Some species quickly retreat to the burrow bottom and remain there. Other species tend to reappear in minutes. In any case, it is difficult to approach most crawfish during daylight to make a catch.

Daytime catching of burrowers is best done after heavy rains. Watching weather forecasts helps to select good times to visit burrower habitats. Rainfall of one to several inches results in standing water for several days in low areas with poor drainage. A two-inch layer above ground is typical. These wet places are favored home sites of burrowers. Following a rain that covers a field riddled with burrows, burrowers usually emerge immediately and venture into the fresh standing water for a short period.

Dip net

Burrows may be found in or near ditches that are normally dry. Good ditches route water of up to 4 inches deep for several hours after heavy rainfall. When these dry ditches begin to flow, burrowers emerge and can be caught by pulling a dip net upstream. A word of caution: avoid approaching flood water that is expected to be more than 6 inches deep following heavy rains. The catch doesn't result in anything and danger, especially near culverts, is sure.

Hand Grab

The burrows of pond and lake species are found along the shore margin. These standing waters fluctuate up and down depending on precipitation. Following a rain that causes the water line to cover burrow openings, these species are usually found the next day busily cleaning out the burrow. Find the active burrows and wait to the rear of the direction to which the crawfish will come out. When the crawfish emerges, be ready to pin it down by hand and simultaneously block the hole to prevent escape.

Baited Lines

Baited lines may be used with some success to catch crawfishes from burrows during daytime. The best type of burrow to select for line catching has an opening with signs of activity and a simple tunnel that is directed vertically downward. The burrow should lie in an open area where it is frequently flooded with rainwater. Bacon or some other bait is attached to the line. A weight is attached to the line so that the bait can be lowered downward to the bottom of the tunnel, or to a place where crawfish has taken a position. The bait is slowly withdrawn until it just reaches the top of the water. If a crawfish follows the bait, its movement can be seen or its claw tips spotted. Retreat is then cut off by some method, such as with a shovel blade.

An adaptation of the above method is to dig a shallow trench about as long as a shoe and 3 to 4 inches deep next to the chosen burrow opening. Carefully remove the partition between the burrow opening and the trench, taking care not to drop soil down the burrow. Fill the hole and trench with water. Stir the trench water so that it becomes muddy. Use the string method to coax the crawfish upward and into the newly dug space. Once the resident is pulled into the horizontal trench, a hand grab or blocking maneuver becomes possible.

Easy crawfish catching by yanking them out of holes with baited lines is a myth. As mentioned above, a crawfish catches food with the claws and immediately passes it backward to its smaller legs for eating. From these, bait is easily separated. Line catching is coaxing at best. Strings tied to sticks work better than hand held.

Digging

Though laborious, digging to the bottom of a burrow is perhaps the most effective way to obtain burrowers. After locating an active burrow to excavate, begin by digging out shallow branches that connect to the main downward tunnel. Once the primary tunnel is determined, continue to remove soil downward along its course. The diameter of the hole should be great enough (at least 1 foot) to allow ease of access. Try not to crush the animal or lose the course of the tunnel as digging proceeds. The bottom may be an arms length or even deeper.

An oozy slurry is often felt at the bottom of the burrow, rather than a pocket of water. Feeling for crawfish at the bottom may result in a thick paste (composed of the ooze and soil crumbles that fall downward during digging) being packed around the animal. Location and removal of the crawfish in this paste is difficult and crushing the animal with the digger's fingers is likely. Adding water to the hole as the digging nears the bottom lets one remove the crawfish from the water instead of mud, and greatly lessens the chance of crushing.

Once the excavation is complete, the hole must be refilled to avoid long term damage to the terrain. Burrow excavations may be placed on a plastic sheet or tarp spread near the digging to enable complete return of soil to the hole when finished.

Devil crawfish are reported to be catchable from burrows that have a surface near the water table. The chimney is removed and a hole is dug 2 to 4 inches away from, but parallel to, the main tunnel. Care is taken not to drop soil down the tunnel. The hole is dug to a level just below the water line. Next the hole is enlarged by dissecting the wall of soil between the dug hole and burrow tunnel, again not letting soil drop into the tunnel and clog it. When a depth is reached where the two holes become one, the water is stirred to make it muddy. The crawfisher then waits. The crawfish may emerge after several minutes to check out the disturbance. Its presence is detected by waving antennae. The water's muddiness keeps the collector from being seen until he can quickly hand pin the crawfish to the side of the hole. Most crawfishes do not readily emerge like the devils when this method is used.

Digging is usually limited to open areas where tree and bush roots are not obstacles. There are some hazards. Hands do a lot of digging even when spades or shovels are used. Especially in roadside areas, broken glass may be encountered in the excavation. On occasion, spiders, snakes and frogs also use the tunnels. In any case, digging crawfishes is not fun.

Nighttime methods

During nights of favorable weather, burrowers usually emerge from deep in the burrow. They typically crawl to the opening of the burrow, their claws and eyes protruding. If they see movement, they quickly drop backward down their hole.

Best nighttime catching is during warm weather. On wet summer nights, burrowers can sometime be picked up by hand as they venture away from the burrow. When they are at the top of the burrows peering out, attempts to hand-pin crawfish can be frustrating because catch success is low and specimen damage is high. Other methods are better. Though warm weather is better for seeking burrowers, they may be found in winter too. We have caught them from the top of burrows when air temperatures were as low as thirty-two degrees Fahrenheit.

Flashlight and shovel

A flashlight and shovel can be a good method to catch crawfish at night if soil conditions are suitable. The soil surrounding the burrow entry should be reasonably free of rocks and thick roots, and the soil itself should be somewhat sandy, but not dense and sticky. The area of the burrows is visited before dark, and active burrows are marked with survey flags. Bigger holes mean bigger crawfish. After dark, the burrows are approached cautiously using the flashlight to spot a crawfish at an opening and to blind it of your movements. While holding the beam on it, one slowly approaches the hole with the shovel. While continuing to hold the beam on the crawfish, use a foot to quickly thrust the shovel into the earth near the hole at an angle that blocks the downward passage beneath the crawfish and thus its escape. Once the maneuver has been completed, the crawfish can be extracted by digging down only a few inches.

Hand Grab and Dip Net

While water stands above an inundated field or roadside ditch, a nice catch of specimens may be taken at night from water of less than 2 inches depth. Return to a location at night where active burrows were marked in daytime. The burrowers are spotted in the shallow clear water using a flashlight and then hand grabbed or dipped. Especially good are roadsides that were recently mowed.

Traps

Two general trap types are useful in nighttime catching of burrowers. One is a pit trap. A hole is dug to fit a plastic beverage cup with smooth inner surfaces. A foot-driven bulb planter will quickly dig holes of an approximate dimension for such cups. A 16-ounce cup is suitable for small species and a 32-ounce size for larger ones. The cup lip must be at, or just below, the soil surface or the animals will go around the trap. Bait, such as cheese or peanut butter, may be added as an attractant. Pit traps are set within a field of active burrows and usually checked the next morning.

A second type is the cylinder or can trap. Here the cylinder is set onto, or forced into, the opening of an active burrow. The bottom or entry end of the device has either a trap door or funnel opening. Trap doors are opened by the crawfish as it ascends and close by gravity.

Holding and Transporting Caught Crawfish

Once captured, one needs a good way to retain the catch and keep the captives alive. Keep them cool, in shallow water, separate and confined. Overheating usually happens when direct sunlight shines on crawfish and their containers, or when crawfish experience the hot conditions of a parked vehicle. Keeping captive crawfish in water that is not deeper that the top of their bodies enables them to tilt to one side and thereby expose a set of gills to air at the water surface. This behavior allows a small gill bailer to move water across the gills and prevent suffocation should the holding water become insufficient in oxygen. In addition, the shallow water permits a greater efficiency of oxygen and carbon dioxide exchange with the air.

Separation into individual containers prevents the injury and death that occur when captives are bunched. A suitable container is one that prevents crawl out. We use wide mouth bottles of 7.5 inches in depth and 3.5 inches in diameter, and pack them 12 per in styrene ice chests. This size is suitable for large to small sized crawfish. If very small crawfish species or small juveniles of larger species are to be retained, the wide mouth bottles are freed for continued use by transferring small catch to smaller bottles of 3.5 by 1.5 inches. The boxed bottles are carried to the catching site where crawfish are added as they are caught. When on trips to catch crawfish at multiple sites, the catch location is immediately written on the bottles with a fine point, permanent ink marker in order to avoid later confusion.

Crawfish should not be retained in transfer bottles for more than two days. Those that enter molting during the period after capture could die in the presence of a buildup of metabolic waste and reduced water oxygen. If they must be kept longer than two days, exchange the water.

Catch and Release

A crawfish catch can be maintained for a while at the point of capture and then released. After a desired period of enjoyment, the crawfish should be returned to the place from which they were taken. Some form of temporary containment is usually appropriate. Otherwise, the crawfish specimens are hand held only for a few seconds prior to release. Transferring crawfish to a holding container allows for good close up examination. Photos may also be taken. Suggestions made above for holding and transporting also apply for these shorter periods of confinement. Plastic shoe boxes or bowls are better for observation than bottles and deep buckets.

Observing Crawdads in the Natural Setting

Daytime crawfish watching is best done from a vantage point some distance above the spot where they are active. This may be a creek bank or the top of a culvert. The water must be clear for good viewing. To keep from being noticed, avoid making abrupt movements while slowly approaching the viewing position.

Interesting observations of crawfish in clear streams and pools can be made at night. A lantern or flashlight that shines a broad beam is best. Crawfish are spotted by the bright, ruby shine that reflects from their eyes. After an initial period of wariness about the light, the crawfish continue with regular activity.

A few crawfish species show a natural curiosity. When they see you, for example, they may crawl onto an underwater rock to try viewing you as you view them.

Planning Crawfishing Trips

Where to go

The assemblage of crawfish species can change from habitat to habitat within a few feet. The general locality, however, will contain only a limited number of species, the list of which is determined by species distribution. An awareness of the kinds of crawfishes that occur in a place helps with identification once they are caught. Knowledge of the habits of each kind helps to pick spots where they more likely will be found and caught.

Knowing the names of creeks and limits of an area's watersheds is useful. These may be learned using county maps produced by the state transportation department, or using the relatively inexpensive books (atlases) that identify roads, streams and other features in a high scale presentation.

Making a pre trip survey of an area for signs of crawfish can make later trips more successful. Look for burrows, burrow activity, shed exoskeletons, crawfish in clear water and spots identifiable as good habitat. Recording such information in a notebook helps to remember good places and to plan future outings. Poor places such as bug holes or places with hazards could also be noted. Survey flags may be used as markers so that exact spots can be found during future visits. Flags are especially helpful for nighttime returns.

Several stops are usual for crawfishing excursions. Selecting these in advance makes a trip plan proceed smoothly and timely.

When to go

Choosing a favorable time for a crawfishing trip usually results in more enjoyment and a greater catch. Milder days in cooler months are usually better for catching most species than hot days in warmer months. The mosquitoes, ticks and chiggers are less bothersome, and snakes are hibernating. Spring months are favorable because new vegetation, both aquatic and terrestrial, has yet to fully establish. During summer and fall, one can seek out crawfishes in mowed roadways and parks. Burrowing species usually remain below ground during dry periods of summer and early fall. This can change quickly, however, when rainstorms come as the result of the hurricane season or other climate changes. Dry season or drought periods provide some advantage for catching stream species. At these times, some streams that would otherwise flow with deep water become shallow enough to enter for successful dipping and hand catching.

Daily activities such as driving or walking are times when one could make general observations about surface water conditions that favor crawfishing. Things to look for are the presence and depth of water in ditches and streams and how active crawfish have become in building and reconstructing burrows. Advanced notice for a successful post rain outing is given by the local weather forecast.

How to go

Take everything that is needed. A list helps. Remember to take a fishing license if one is required. Choose equipment suitable for the required catching methods. Going by foot or by vehicle will determine whether one will have to travel light or not.

Footwear is important. Knee boots are the most versatile. They may be carried until the site is reached. They are also easily removed to prevent transfer of mud to a vehicle when moving from place to place. Seining in summer may be done while wearing sneakers. Hip or chest waders are nice in cold months or in spring fed streams.

Catching crawfish is more enjoyable if it is done with one or two other people. Members of the party can work close together or spread out. Short range walkie-talkies can be used to communicate when people are out of an earshot.

Cautions

Remember to stay away from swift water that flows in ditches and creeks after rains. It is unlikely that crawfish would be caught in such water and falling into such water is very dangerous. On such occasions, restrict crawfishing to lowlands and drainage ditches where water is less than 6 inches deep. Remember also to avoid creeks and rivers that have swift flow. Being caught up in such water with boots on is not a happy event. Some consideration should be given to catching crawfish in riffle areas where the possibility exists of a sudden downstream rush of water from flash floods or periodic reservoir release.

Cautions about crawfishing around deep water are much the same as one uses when not crawfishing. Be careful about falling in deep water and getting too close to drop offs. Drop offs have a steep grade with a bottom that is firm and slick. If the bottom cannot be seen, make an initial check with a pole.

Along roadsides and around bridges, surface water and shoreline may have an abundance of sharp items such as broken glass, barbed wire, re bar, concrete and can lids. Awareness of the possibility of these items will avoid many accidents. Utility lines are also common in these areas. Signs may be posted that warn of hazards and the location of crossings and underground routes.

Slick rocks are usually present where sufficient nutrients exist in the water and the submerged rocks are exposed to sunlight. A banana peel fall can be painful. Consider if walking across such surfaces is worth the risk.

Having both feet stuck in muck can be frightening and cause one to wonder if the boots will be lost. Check bottoms with a pole before venturing into unfamiliar water or over a soft soil. Becoming stuck can be dangerous in some places such as the silty stream-side flats of the Red River.

Getting to water in some cases is just not worth the effort. Briars, brush, vines (including poison ivy) and steep grades usually spell misery. Negotiating a way through such stuff lessens awareness. Holes in the ground or between rocks can result in ankle or leg injury. Go elsewhere.

Always look and listen for snakes when you arrive at a spot. When they see you, snakes usually take the opportunity to depart. If they do not leave, let them have the place and go somewhere else. Snapping turtles are sometime caught while seining. Their name is a good one. The neck will extend a long distance backward -- long enough to bite someone holding the turtle by the shell. In some coastal areas, alligators are large in population. Gators are sometimes aggressive and dangerous. Avoid crawfishing near their dens.

Mosquitoes, ticks, chiggers and flies can be discouraged with repellant. Wasp nests overhanging the water or path should be watched for. Dip nets and seines catch a variety of water bugs. Some bite. Knowing which kind not to handle is helpful.

Walking around parked vehicles and crossing roads is very dangerous if one is not watchful and does not take the necessary precaution. Be careful not to put children at risk. Park vehicles well off the road shoulders for unloading gear. For example, park in a pull-off, unused field access or parking lot.

Consider not crawfishing far away from your parked vehicle. Theft may take place if thieves can see you far away from your property. Avoid places where homeless people, target shooters and trouble-makers tend to congregate, such as under the bridges of some rivers.

If crawfishing in a stream that crosses a road and flows from, or onto, private property, you will be trespassing if you wander onto the property in search of crawfish. The same could be said of flooded areas or burrow fields. In many cases, merely asking permission to crawfish on the property will be met with acceptance and even help. Permission removes the tension and illegal activity and makes the crawfishing more enjoyable.

The annoyance of dogs is often experienced when crawfishing in a ditch or stream crossing that is near to a dwelling. In some cases, the dog could bite. The best bet is to go elsewhere.

Talkin' Crawdad

abandoned – an open burrow with a very dry opening that has no signs of use.
active – an open burrow with fresh, moist subsoil around opening
"big enough to stick a fist in" – a saying that indicates the burrow of a large crawfish
bug hole – a surface water pool or puddle (usually with hot water in daytime) without crawfish or fish
bull, old bull – an exceptionally large male crawfish
"*burrows*" – an evaluation, means crawfish are present
big mama – an exceptionally large female crawfish
cooned out – a ditch or pool without crawfish and an abundance of raccoon tracks around its edges
crawdadder – a person who goes crawdadding or crawfishing, crawfisher means the same thing
Crawdaddy – a designation meaning that one has reached an esteemed status due to long experience or exceptional exploits
crawdaddy -- a term equal in meaning to crawdad and crawfish.
crawl under, a crawl under – wandering on foot with a dip net through a large, under-road culvert containing water of less than 2 inches deep; must be bright enough to spot snakes, wasp nests, etc.
"dry!" – a call of a passenger that designates the status of a stream or roadside ditch while traveling at more than fifty mph
dump – to empty the bulging contents of a dip net onto land at a relatively clean smooth spot for visual inspection of critter movement
"*enough to float a simulans*" – a large amount of rainfall in a short period of time; days following such an event makes for good catching of grays
fish hole – a surface water pool or puddle without crawfish but with large fish present
flip – when a dip net or seine is turned upside down for removal of unwanted material
fresh molt – a live crawfish that has recently molted
"*good'n*" – what one should say when a crawfishing partner walks up and presents a large perfect crawdad with a smug on his or her face
hittin' – On trips where multiple stops are made, a stop is a hit. Hittin' is the process.
in berry – a female crawfish with eggs or young attached
juvie – a small, immature crawfish
"Kin I hep ya?" – saying of an irritating person who really doesn't want to help but rather is just curious and wants to talk; *'Watcha doin?"* means the same thing.
"*lookin' good*" – an evaluation of a place given as its perception unfolds
mess – mud from seines, boots or nets spread here and there inside a vehicle
miss – one that got away
molt – the shed exoskeleton noticed in the water or on land
"pass?" – what one person asks another about minimal value of a crawfishing spot
pop – to strenuously shake a dip net or seine (in unison) to remove excess water
plugged – a burrow opening that has been sealed with mud
"ready?" – what one person asks another about whether to continue crawfishing at a spot
regrowth – a crawfish with a claw that has replaced the original
rinse – prior to departure from a site, a quick water cleaning of a mud laden seine or dip net
soft – status of a bottom too soft for good wading
snag – something underwater that hangs a seine
"stuck" – what someone says when feet or boots have sunk too deep to lift out
sweeping – working shallow water of a flooded flatland with a dip net after a heavy rain
"swing!" (around) – what one seiner says to the other when it is time to pull into shore
"thick" – what one person says to another when evaluating a spot that has too much undesirable vegetation to pass easily from where you are to the water
trash – the litter or vegetation collected in a net in addition to crawfish. Crawfish are usually found beneath loose trash rather than within it
"up!" – what one seiner says to the other when it is time to lift the seine directly from the water
"water!" – a call of a passenger that designates the status of a stream or roadside ditch while traveling at more than fifty mph
"yaaa!," "yeow!," "oww!"– said by someone being pinched

What's that? Animals commonly observed when catching crawdads

On the crawdad shell

The most common animal noticed on crawfish shell surfaces is a tiny leech-like organism called a branchiobdellid worm. Branchiobdellids are about the diameter of a sewing thread and stretch to about 0.25 inch. They have a cream color. Suckers on the head and the tail help them move in inch worm fashion. Branchiobdellids deposit eggs on the crawfish shell. Eggs are most often noticed on the top of the rostrum. Aquatic bugs also lay eggs on the surface of crawfish. These eggs are several times larger than those of the branchiobdellids.

Branchiobdellids on the surface of a crawfish. Worm and eggs.

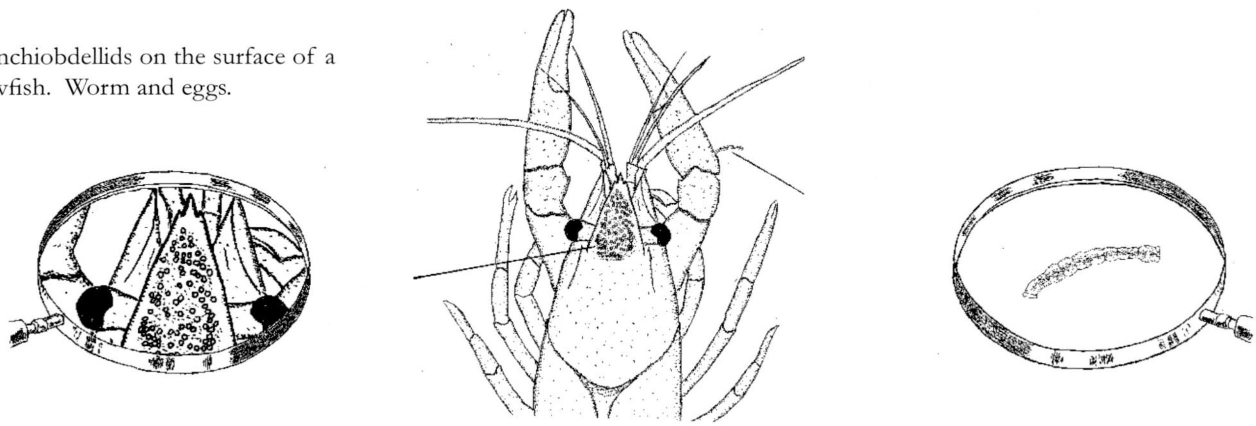

In a net catch

Insects

Every crawfishing trip results also in a catch of a variety of aquatic insects. Insects have a body that is divided in primitive form into head, thorax and abdomen. They have two pairs of wings and three pairs of jointed legs on their thorax. The second wing pair may be reduced as in adult mosquitoes, or comparable to the first as in dragonflies.

Adult and sub-adult aquatic insects are caught along with crawfish when using a net or seine. Most insect catch consists of sub-adults (the synonymous terms given for the sub-adult form of an aquatic insect are larva, nymph and naiad). These feed on animals smaller than themselves, soil litter and plant life. Adult forms have a diversity of diets but most of the ones seen in net catches are predators. Larger adult and sub-adult insects (0.5 inch or more in body length) are illustrated below.

Common adult insect catches:

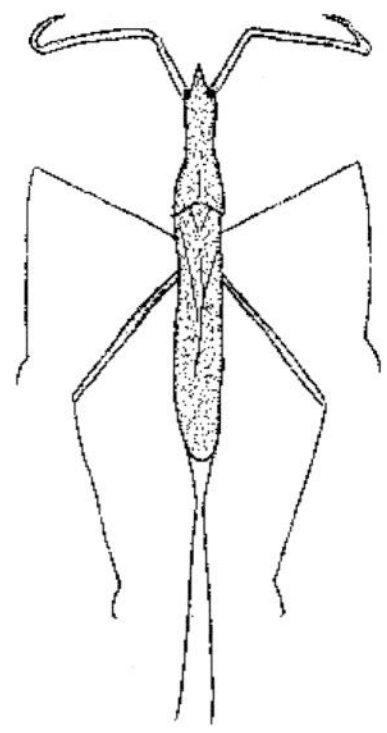

Water scorpions. Crawl on vegetation and bottom litter. They are brownish in color and have forward appendages adapted for grasping prey. Water scorpions, like other predaceous bugs, have a piercing mouthpart that they use to subdue prey. Water scorpions bite people. They are usually noticed as adults and grow to 1.5 inch long.

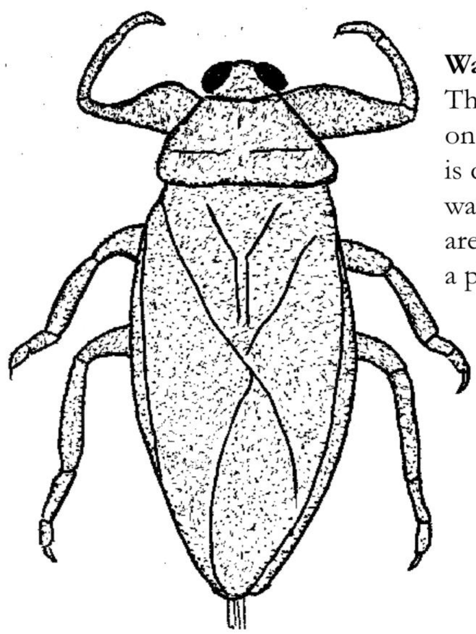

Water bugs. Giant water bugs consist of larger (> 1 inch) and smaller types. The common smaller types have wider abdomens. Some water bugs lay eggs on the back of the males. For those that do, finding egg carriers among them is common. Giant water bugs swim but spend most of the time crawling under water or waiting for some food to swim by. Some are good fliers. Their eggs are occasionally found attached to the surface of a crawfish. These insects give a painful bite to people.

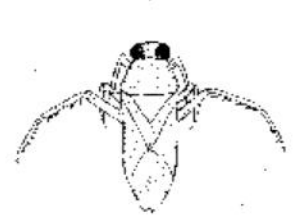

Backswimmers and water boatmen. Backswimmers and water boatmen have a very similar appearance to each other. They are seen at the water surface but quickly dive when approached. The topside of a backswimmer is keel shaped. Water boatmen lay eggs on crawfish. Backswimmers and water boatmen are about 0.5 inch long. Backswimmers bite people.

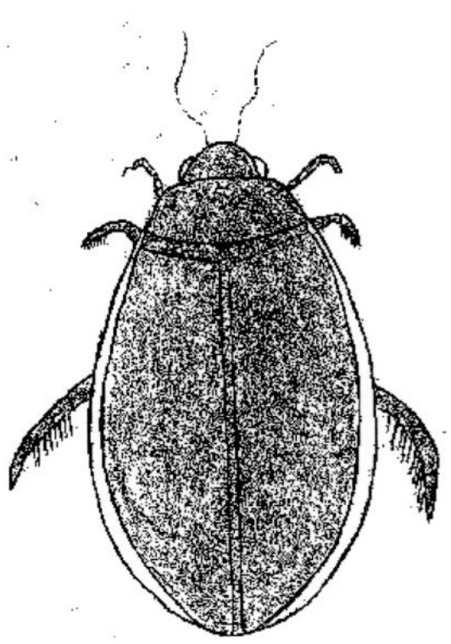

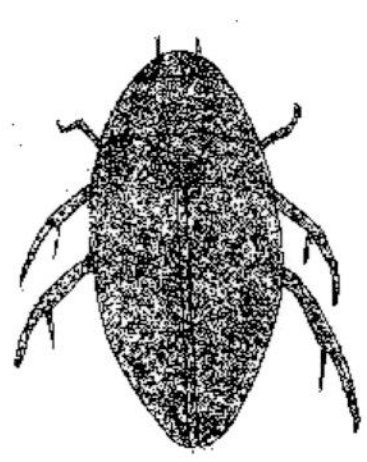

Beetles. Most beetles are land animals but many also live in and around water. Diving beetles (1.5 inch) look somewhat like the also dark colored scavenger beetles (1 inch) but the sides of diving beetles bear light stripes. These beetles make contact with the surface for air and dive for food. Diving beetles surface tail up while scavenger beetles surface head up. Smaller whirligig beetles (0.5 inch) dive from the water surface, swim rapidly and spin around. They are more often seen than caught.

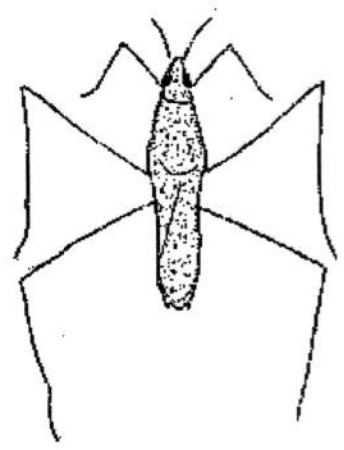

Water striders. These are also more often seen than caught. Water striders (0.75 inch) jump and skate across the water surface.

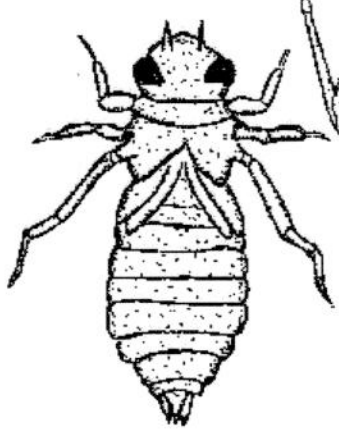
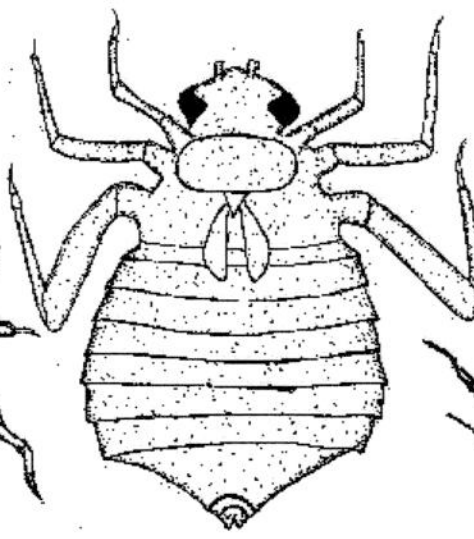
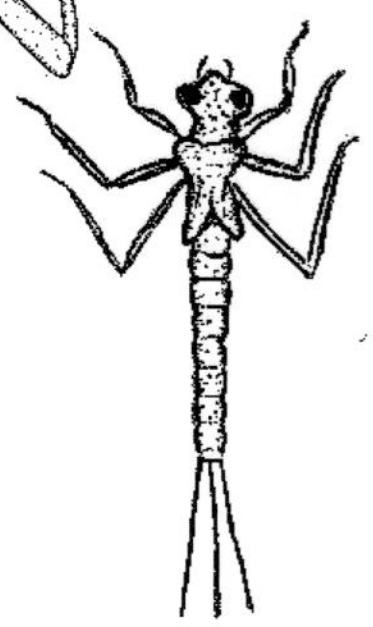

Common larval insect catches:

Dragonflies and damselflies. Damselfly nymphs (0.5 to 1 inch) are slender with three leaf-like structures extending from the tail. Dragonfly larvae (0.5 to 2 inches) have stockier bodies than damsel larvae and no tail extensions. The abdomens of dragonflies enlarge with age and body shape may be dissimilar between species.

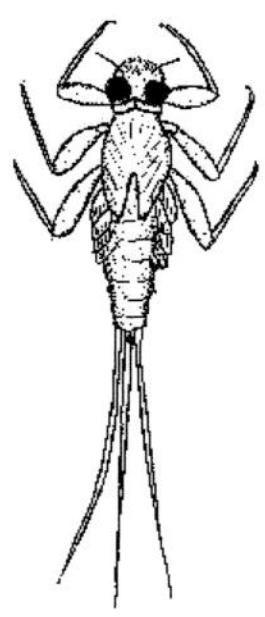
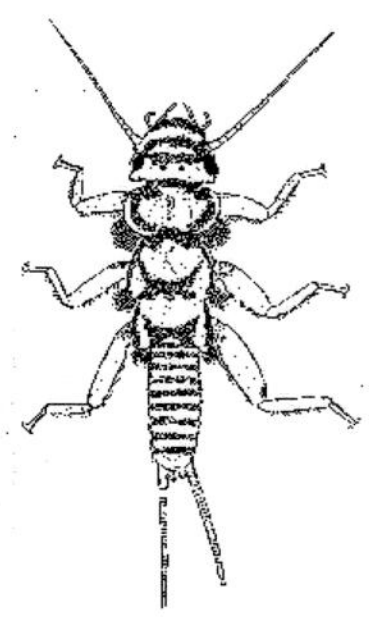

Mayflies and stoneflies. Nymphs of mayflies (1 inch) and stoneflies (1 to 2 inches) are common in streams. Mayflies have leaf-like gills extending from the sides of the abdomen; stoneflies have filamentous gills that extend from behind the legs. The general appearance of both mayfly and stonefly larvae varies strongly between species and with age.

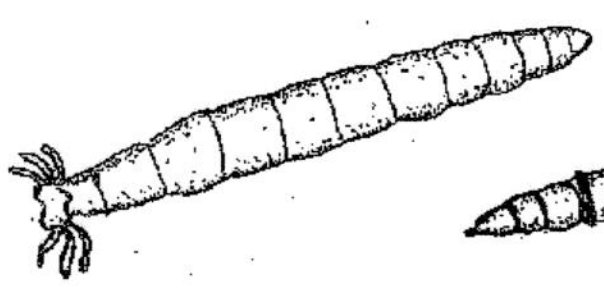

Craneflies and horseflies. Cranefly larvae and horsefly larvae look like robust, legless worms or elongated maggots. The larva of a cranefly has a disk on the posterior end that has some arrangement of gills and "breathing" tubes. A horsefly larva is usually smaller (0.5 to 1.5 inch) than a cranefly (1.5 inch) larva.

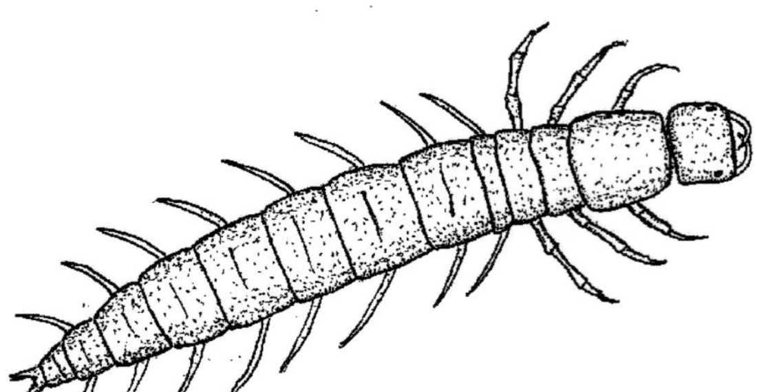

Dobsonflies. The larvae of dobsonflies, called hellgrammites, have broad square heads with pincers. They are aggressive and will pinch fingers if not handled properly. They grow to more than 3 inches depending on age and species.

Beetles. Aquatic larvae of beetles are also caught by net. In general, they resemble beetle larvae (grubs) dug from lawn soil but are smaller and more elongated.

Spiders

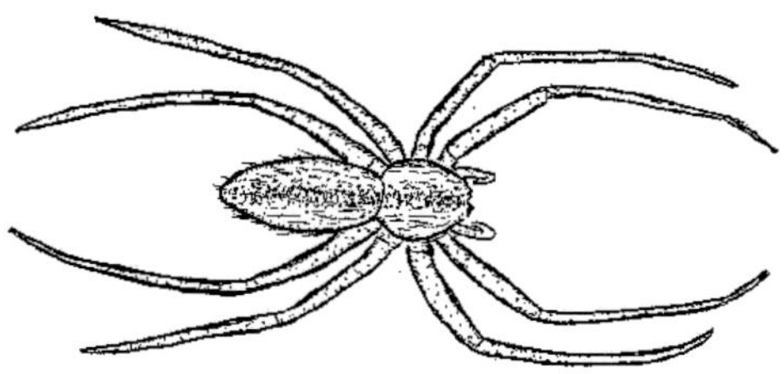

Spiders (semi-aquatic or fishing spiders, 0.5 to 0.75 inch body length) are often seen walking about on the water surface. They catch small aquatic animals and dive when disturbed. When caught in a net, they quickly scamper out. They are able to bite a human, but they rarely do.

Snails and mussels

Aquatic snails and clams, called mussels, are frequently scooped up along with crawfish. The presence of mussels is indicated by the shells of dead animals that are noticed in shallows or along a shore.

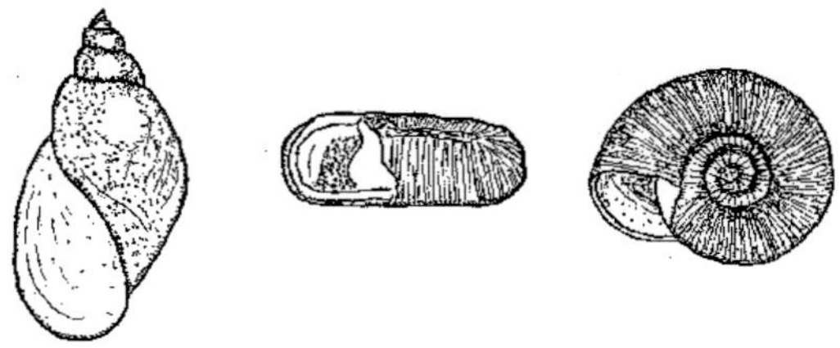

Snails. Snails crawl and eat plants and dead materials. When disturbed, they retract into their shell. As the snail grows, the shell enlarges as a whorl and the opening increases in diameter. The overall shell shape is that of a cone, or, in the case where the whorl expands outward instead of downward, a flattened coil. The aquatic snail shell is colored tan, gray, green or black on its surface and is composed mostly of lime secreted by the soft body. The greatest length of most freshwater snail shells is less than 1 inch and usually less than 0.5 inch. Eggs are laid on twigs and other solid structure where they develop in transparent gel-like masses about the size of a pencil eraser.

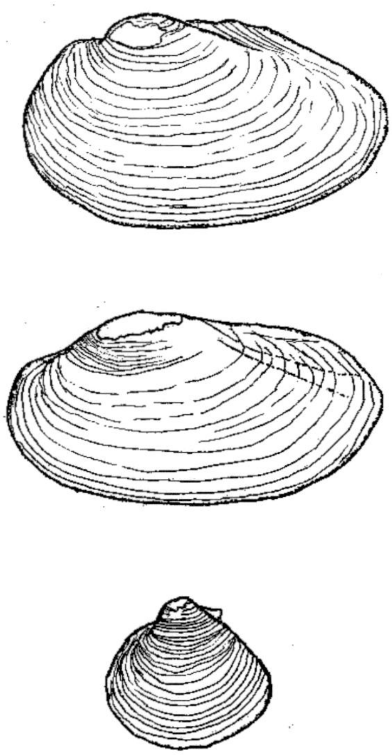

Mussels (freshwater clams). The mussel shell is composed of two valves hinged together on one side. A soft body with gills lies inside. Shell valves close tightly and open slightly by columnar-shaped muscles that stretch from one valve inner side to the other. Depending on the species, the shell's outer surface is brown, yellow, green or black and the inner surface is usually pearly white or cream, but sometimes purple or pink. Shell shape is oval or round in side view and but in frontal view it is thickly or flatly oval. Shell length may reach 10 inches in one species but most reach less than 5 inches. A common round-shelled species grows to the size of a thumbnail.

Mussels live burrowed in the top layer of the bottom with perhaps 90 percent of their shell and body beneath the surface. Mussels move slowly in the bottom soil with the help of a muscular extension called the foot. The valves open slightly to allow water and detritus (food) to enter and leave. The opening also serves as a portal for oxygenated water to reach the gills and waste to exit. Sperm also enter through the opening during fertilization. When larvae develop sufficiently, they are released into the water where they attach to the gills and skin of fish and live for awhile as parasites. After further development, they release and fall to the bottom as small mussels.

Worms and leeches

Worms and leeches are sometime noticed upon careful observation of the litter. The bodies of aquatic worms, like leeches, are made of a series of segments that are apparent in a closer look. Larger, more noticeable worms are long and stringy like earthworms. Smaller, shorter (< 0.25 inch) kinds are rarely noticeable and usually have bristles (setae) extending from the sides of their bodies.

Leeches. Parasitic can be seen attached to a person's skin or boots when wading out of the water. The leeches that attach (rarely) to people are mostly fish or turtle parasites. Typically, these have flattened bodies. Some smaller, tubular-shaped leeches are good swimmers and are predators of small aquatic animals, instead of parasites of large ones. These are commonly brownish in color and contrast with the darker shades of the litter. They are 0.5 to 1 inch long.

Fishes

A variety of fish common to shallow waters where crawfish are caught may be netted while crawfishing. A captured fish is usually one that is small as an adult or a smaller stage of a larger fish species. Commonly caught full-grown fish are mosquito fish, madtom, pirate perch, pygmy sunfish, top minnow, darters and various minnows. A good list of juvenile fish includes black bullhead, channel cat, largemouth bass, crappie, carp, shad, bowfin, pickerel, various minnows and various sunfishes.

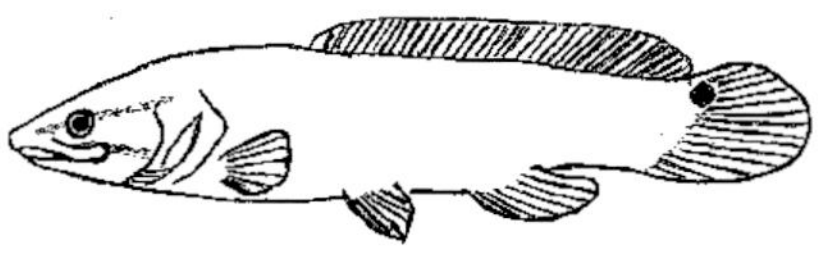

Bowfin (grindle, dogfish). This fish lives in south-central and east Texas. Males have a spot at the base of the tail fin. Young are occasionally caught while crawfishing. Adults are five or more pounds.

Longnose gar. The longnose gar has a statewide distribution. Of the four Texas gar (longnose, spotted, alligator and shortnose), the longnose is most likely caught, followed by the spotted. Gars have long, cylindrical bodies, a head covered with bony plates and elongated jaws lined with many sharp teeth. They periodically top to exchange the air in their air bladders. Gars reach three to 5 feet except for the alligator gar which may grow to 12 feet in length.

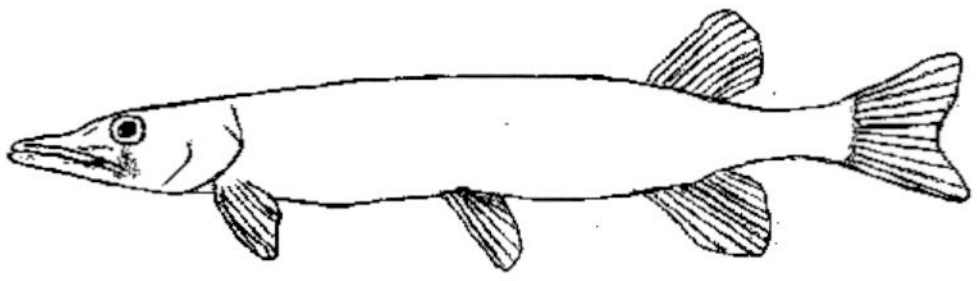

Redfin pickerel. This fish has an East Texas distribution. Adults, usually less than 1 foot long but juveniles (<½ ft) are most likely caught while crawfishing. The other Texas pickerel (the chain pickerel) lives in northeast Texas and is less commonly caught. It grows to more than 2 feet long.

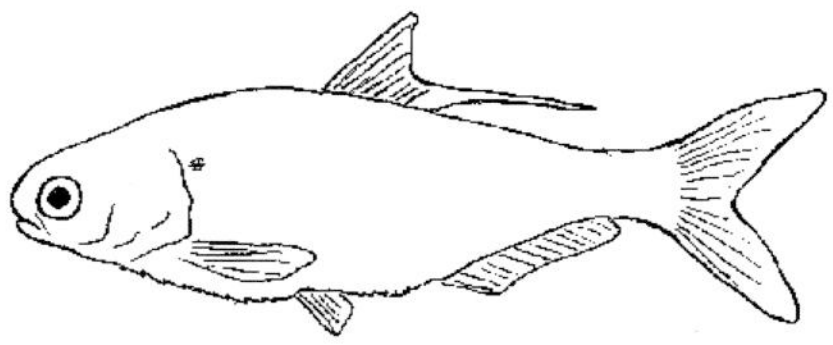

Gizzard shad. This shad has a statewide distribution. It has a flat body with a saw-toothed edge on the belly. Adults to 1.5 foot. The other Texas shad (threadfin) is unlikely to be caught when crawfishing.

Carp. The common carp has a statewide distribution. The adults commonly reach 30 inches but can grow larger. Crawfishing catches are less than 3 inches long. An obvious barbel protrudes outward and downward from the back of the jaw on each side.

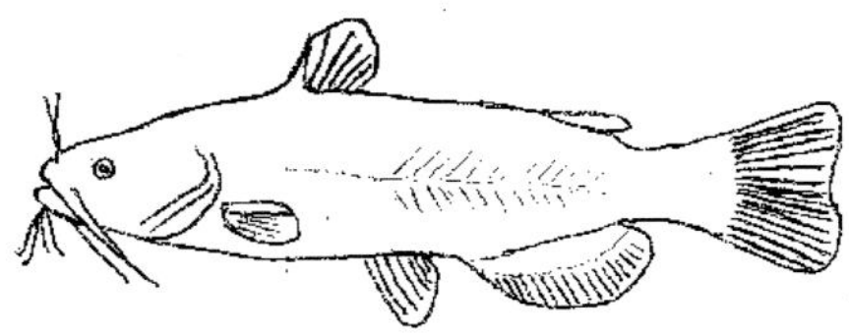

Black bullhead. This fish has a statewide distribution and is more commonly encountered than the other Texas bullhead (the yellow bullhead). The black bullhead is distinguished by having black chin whiskers and a slightly forked tail. Black bullhead adults are commonly 1 foot long but grow larger. Crawfishing catches are common, usually of 3 inches or less. Those less than 1 inch swim as a school in shallow water. These fish schools appear as dark spheres, 1 foot in diameter or less, and make one of the most curious sights of a crawfishing trip.

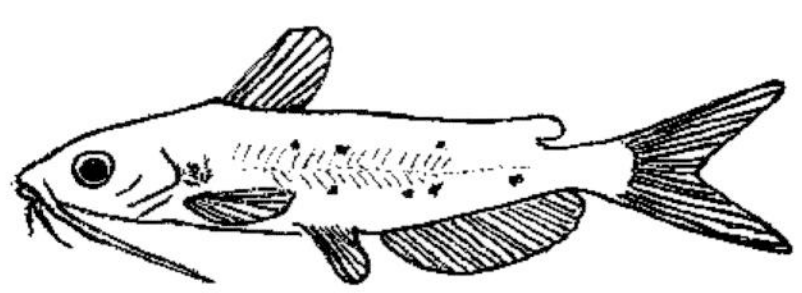

Channel catfish. The channel cat has a statewide distribution. It is commonly caught with seine and dip net at sizes of 2 to 5 inches. The other Texas catfish with a deeply-forked tail (the blue catfish) is not expected in a crawfish catch

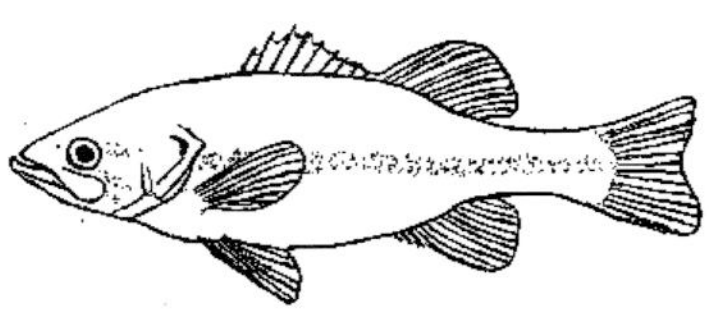

Largemouth bass. The largemouth is very common and is caught statewide. It has a distinctive large mouth and lateral stripe. The young of two other Texas bass (spotted and Guadalupe) could be caught in a favorable habitat (running water). They look almost identical to the largemouth. The introduced smallmouth bass is unlikely to be caught while crawfishing.

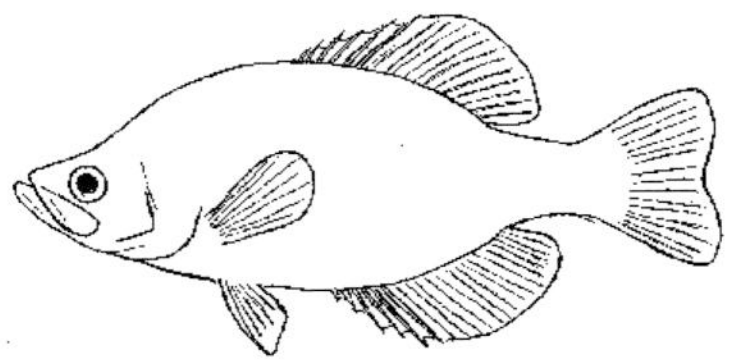

White crappie (white perch). This crappie lives statewide except far west Texas. Its young are occasionally caught in backwaters with a dip net or seine. Adults grow to more than 2 pounds. The other, slightly smaller crappie (the black crappie) is found mostly in the open waters of east and east-central Texas and is less often caught while crawfishing.

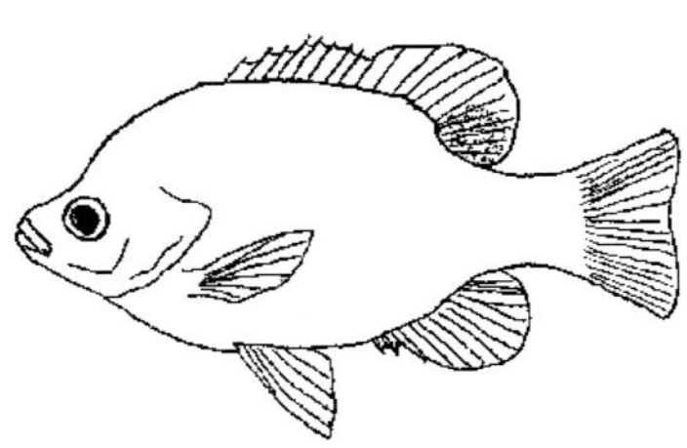

Bluegill (bream). The bluegill has a statewide distribution. It is the most commonly caught of eight Texas sunfish (longear, orangespotted, redear, spotted, bantam, redbreast, dollar) with relatively small mouths and side-to-side narrow bodies. A black spot is present at the rear of top fin. Crawfishing catches are mostly 2 inches or less. The adults reach 10 inches in favorable habitats.

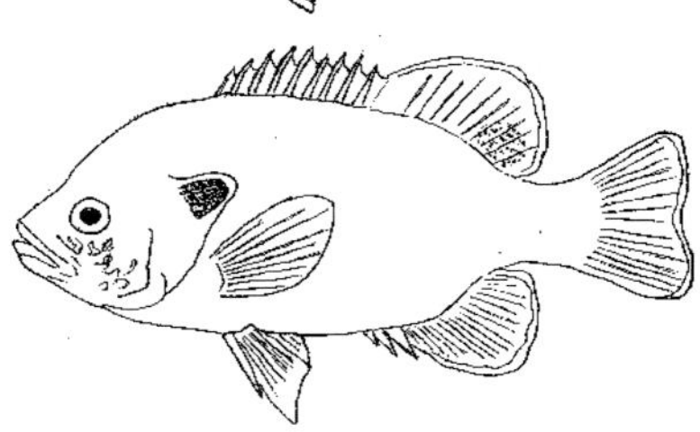

Green sunfish (perch). The green sunfish has a statewide distribution. Though a number of other fishes are also called perch, this fish enjoys a more universal attachment to the name. Its body thickens as it gets older and it has a medium-large mouth. The top fin has a black spot. The adults are 6 inches and smaller.

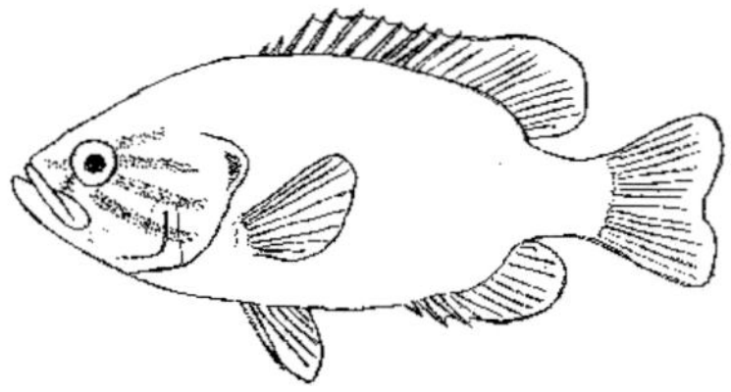

Warmouth (goggle eye). The warmouth occurs statewide. It has a large mouth and a body that thickens with age. Stripes radiate from its eyes. The adults grow to about 10 inches.

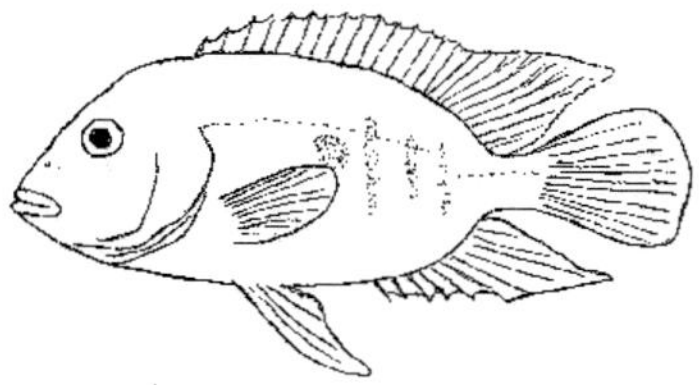

Rio Grande perch. The Rio Grande perch lives in south Texas (Rio Grande drainage) and spring fed streams of west-central Texas. Small ones are commonly caught while crawfishing in south Texas. Large adults are usually less than 8 inches long.

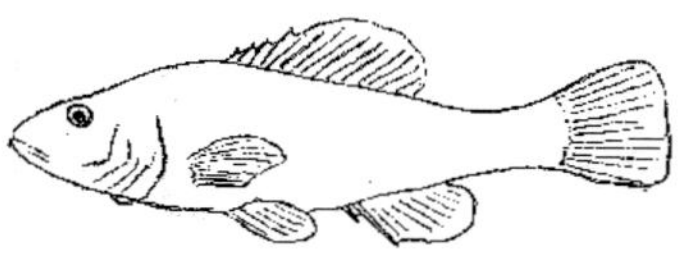

Pirate perch. The distribution of the pirate perch is eastern Texas. Commonly caught while crawfishing in its range, the darkly colored pirate perch has the unusual feature of the anus being positioned forward under the head instead of in front of the anal fin (the rear bottom fin).

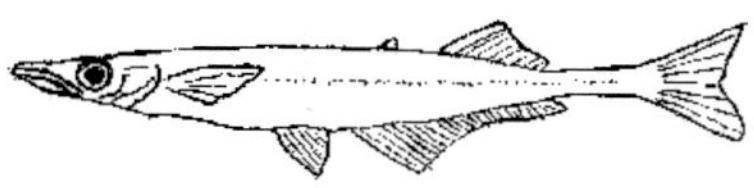

Brook silverside. The distribution of the brook silverside is eastern Texas. The other Texas silverside, the tidewater silverside, has a coastal distribution, but also occurs inland in the Red River and Rio Grande. Silversides have a slender body and a top fin divided by a space.

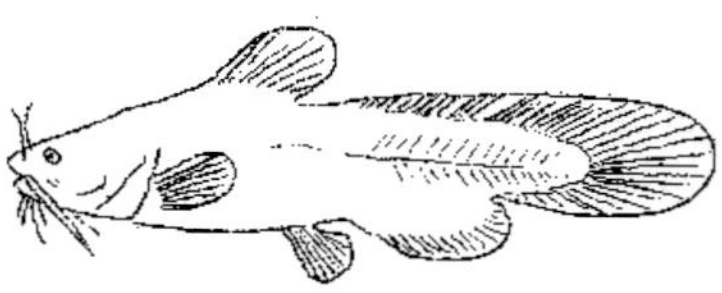

Tadpole madtom. This madtom lives in central and eastern Texas. The other Texas madtom (freckled) is less common and more eastern in distribution. Madtoms are a small species and distinguished from other catfish by having a tail fin that continues up the back and lacking a small, fleshy fin before the tail. The adults grow to 5 inches. The usual crawfishing catch is less than 2 inches.

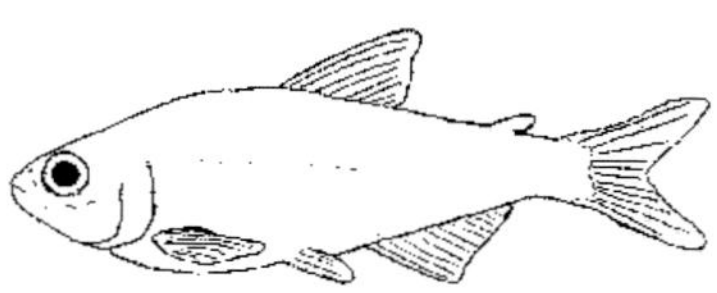

Mexican tetra. The native range of the Mexican in Texas is the lower Rio Grande, Pecos and Nueces river drainages. It has been introduced to several Gulf drainages of central Texas and the Red River in the vicinity of LakeTexoma. A distinctive fleshy fin is located on top and forward of the tail. The maximum size is 4 inches.

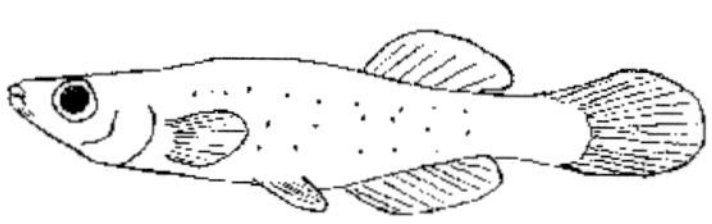

Golden topminnow. The golden topminnow, and the similar Blair's starhead topminnow, are commonly found in clear waters of eastern Texas. They are difficult to catch where water is broad enough for them to swim away. Maximum size is 2.5 inches, but the usual catch is 1.5 inch.

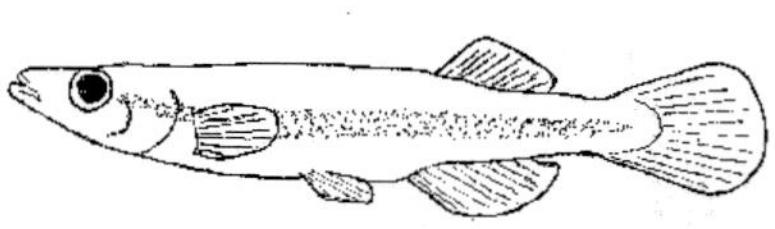

Blackstripe topminnow. The distribution of the blackstripe topminnow is eastern and central Texas. This topminnow and the similar blackspotted topminnow have the distinctive longitudinal stripe and long cylindrical body. Like the golden and starhead topminnows, they swim at the water surface. The adults reach 3.5 inches. Smaller sizes are caught while crawfishing.

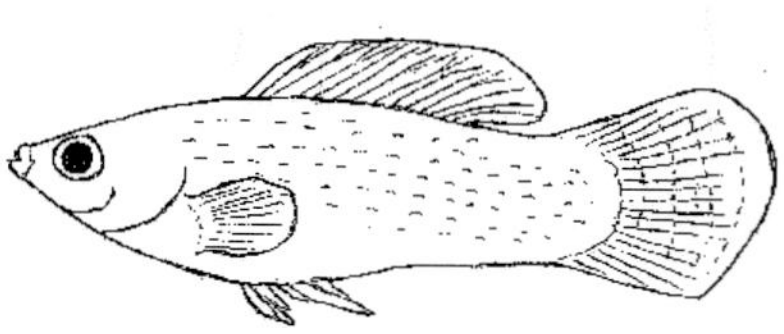

Sailfin molly. The sailfin molly lives in streams and ditches of the flat coastal plain. It is not an egg layer like most other fishes. The young swim on their own immediately after release from female parent. Though usually caught in as a smaller size, the sailfin molly grows to 3 inches.

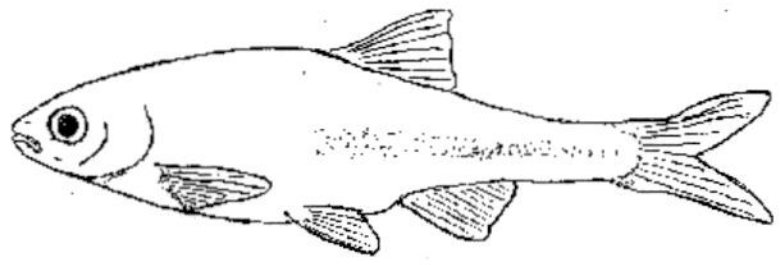

Red shiner. The red shiner has a statewide distribution. The usual catches are 2.5 inches long or smaller. Many minnow species occur in Texas. Some are widespread in the state; others have a more limited range.

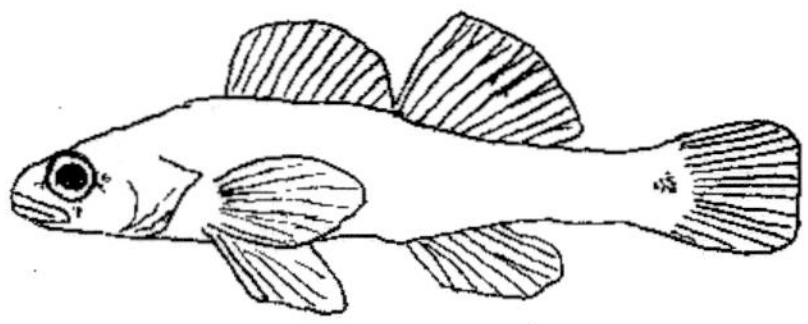

Slough darter. This darter lives in the eastern half of state. The slough darter is often caught in wooded backwaters and ditches. Of nearly twenty Texas darters, half are fairly widespread. Most darters favor flowing water where they hug the bottom and move quickly from here to there. The adults are 2 inches long.

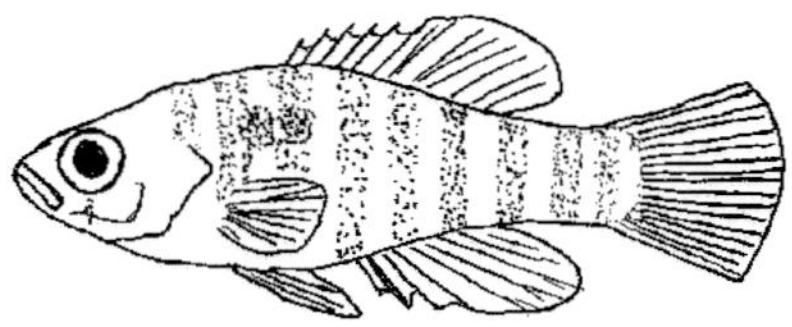

Banded pygmy sunfish. This sunfish lives is eastern Texas. It is the smallest sunfish, growing to only 1.5 inch long. The dip net catches usually fish that are 1 inch or smaller.

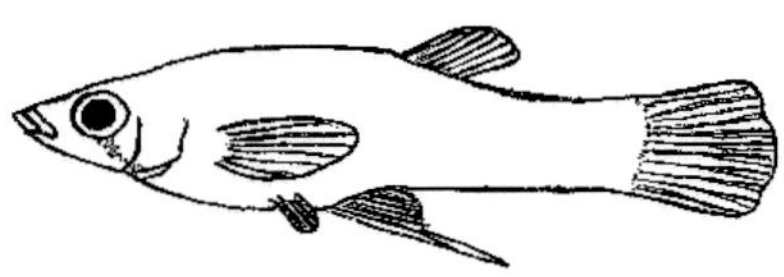

Mosquitofish *(Gambusia)*. The mosquitofish has a statewide distribution. It is another fish that releases young as free swimmers instead of laying eggs. Abundant along shorelines and in shallows, the mosquitofish grows to 1.5 inch. Other closely related and similar looking gambusias occur in specific localities in the Texas Hill Country and west Texas.

Leopard frog

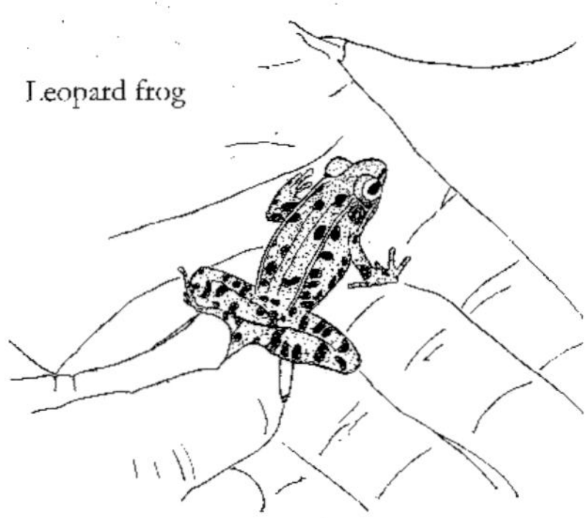

Amphibians

Amphibians are almost always seen or caught on a crawfishing trip. Most have a life cycle that begins with the parent releasing eggs into water. These eggs later hatch into larvae. In many species such as frogs, toads and many salamanders, the larva eventually transforms into a distinctive adult form that remains aquatic (i.e., a bull frog) or ventures onto land, (i.e., some salamanders, toads) later to return for breeding and release of eggs into water. The egg masses are sometime caught in nets or may be noticed along the shores in shallow water in laying seasons. At a closer look, the developing frog, toad or salamander can be seen in each egg within the mass. Except for tadpoles, frogs and salamanders are carnivores.

Bull frog

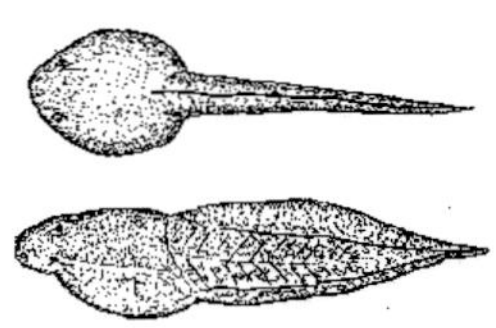

Tadpole

Frogs. Frog tadpoles are the most common amphibian caught during a crawfishing trip. Occasionally, an adult frog will accidentally be scooped up from its temporary hiding place in bottom litter. The bull frog (up to 8 inches long and the largest frog in North America) and the leopard frog are common adult catches.

Salamanders. An aquatic stage salamander is occasionally caught. Typical salamanders have four legs, soft skin, and a long tail. Salamanders laying eggs in water begin larval life with external gills. Kinds that leave water for land life lose the gills upon transformation. Kinds that complete life in water retain the gills. Some salamanders do not lay eggs in water, depositing them instead under logs or in moist soil.

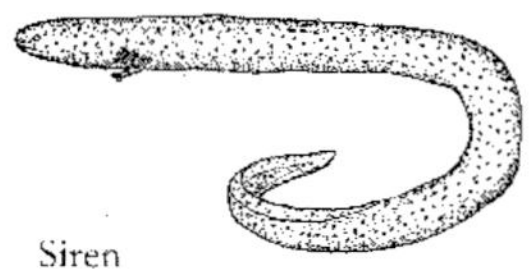

Siren

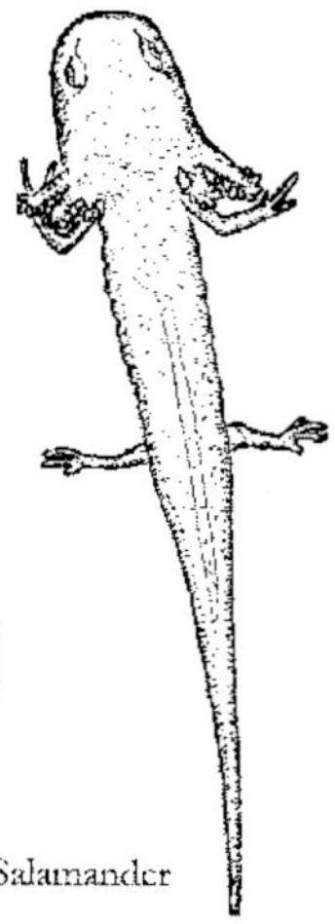

Salamander

Uncommonly, a siren may be caught. Sirens are amphibians that do not transform into a gill-less stage as they become adults. They are long and eel-like in appearance and have two legs forward on the body near the gills. The Texas distribution is eastern Texas and the flat coastal plain to the Rio Grande. Sirens grow to a length of 16 inches. If caught with a dip net , a siren is usually less than 8 inches long.

Rarely, a catch is made of a congo "eel" (*Amphiuma*). A congo eel might be caught in shallow water or from a crawfish burrow in eastern Texas. These eel-like salamanders have four tiny legs and a more elongated shape than sirens. They also grow larger than sirens, to more than 40 inches. They bite when handled.

Turtles

Turtles are more often seen than caught. When caught, the catch is more frequently in a seine than a dip net. The top or back of a turtle is covered with a shell or carapace. In hard shell turtles, the carapace is composed of fused portions called scutes. The belly is also covered with a hard, shell. Hard shell turtles living on land are usually called tortoises, while the aquatic ones are called terrapins. Turtles are egg layers. Aquatic species are largely carnivorous and may scavenge. They are often seen basking in the sun. Crawfish are a favorite food for most aquatic turtles.

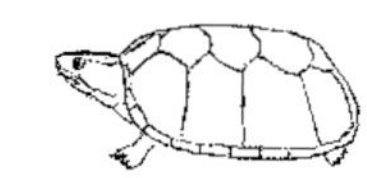

Stinkpot

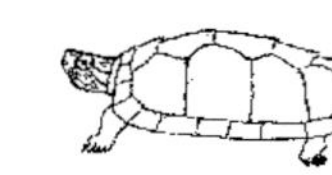

Slider

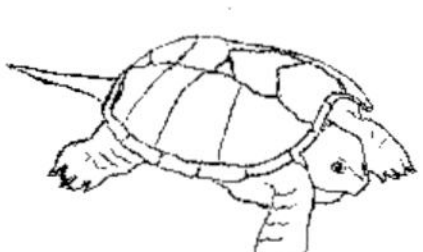

Snapper

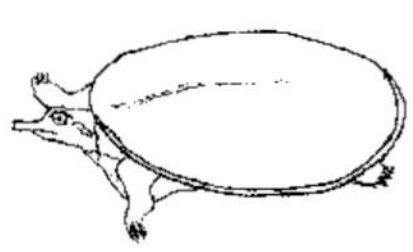

Softshell

Turtles, likely catches. Perhaps the most common turtles that are caught are the mud turtles and musk or stinkpot turtles. Next in frequency are small red ear sliders, softshells and snappers. Eleven marginal scutes on each side of stinkpot and mud turtles separate them from other aquatic hard shells. The other hard shells have twelve scutes per side. The stinkpot has a higher dome and a more elongated carapace than a mud turtle. Snappers are recognized by their large head and tail. Softshells have a pointed snout and a flat, soft and nearly round top shell that lacks scutes. Young red ears have an easily noticed red or orangish stripe behind the eye. Some aggressive aquatic turtles (softshells, snappers and stinkpot) have long necks that can extend quickly to reach as far as their hind legs. Many a careless handler has experienced their painful bite.

Snakes

Snakes rarely end up in netting and if they do it is only long enough to get their bearings for a quick escape. All snakes are carnivores. Some species lay eggs while others release moving, breathing young snakes. The snake species that live in aquatic habitat are mostly of the latter type. Garter snakes and water snakes are most commonly seen around water while crawfishing in Texas. The aquatic snake to be wary of is the cottonmouth, a darkly colored and stout bodied snake that has a poisonous bite. Its head, which is obviously broader than its neck, is held above water as it swims. Another viper called the copperhead occasionally visits shorelines. It and the cottonmouth will hold their ground when disturbed. If a viper is seen, it is time to leave. Vipers have special features like a pit between nostril and eye, fangs and vertical pupils that are distinctive, but field recognition is usually made by observation their broad head and stocky body.

Bryozoans and sponges

Bryozoans and sponges are strange-looking animals that occasionally show up in a net catch. They may also be seen from the water's edge if the water is clear.

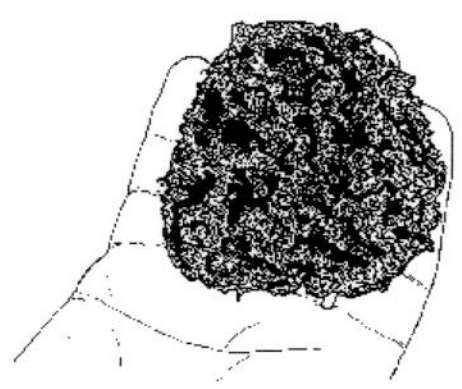

Sponge (22%)

Sponges. Almost all sponges are marine animals. A few kinds live in freshwater. Their appearance is similar to a marine sponge that is commonly sold in stores as a washing aid. Freshwater sponges attach to twigs and litter. They usually have a brownish color though other lighter or darker colors occur.

Bryozoans. Bryozoans are colonial animals that expand as a network of branches or mats. The kind that is most noticeable to crawfishers produces a large roundish or globular mass that may reach the size of a football. The gelatinous mass is firm to the touch, tan colored and, to some degree, transparent. The mass is easily broken apart. Bryozoans that do not produce gelatinous masses usually go unnoticed. These produce a branching network of straw-size tubules that adhere to solid objects.

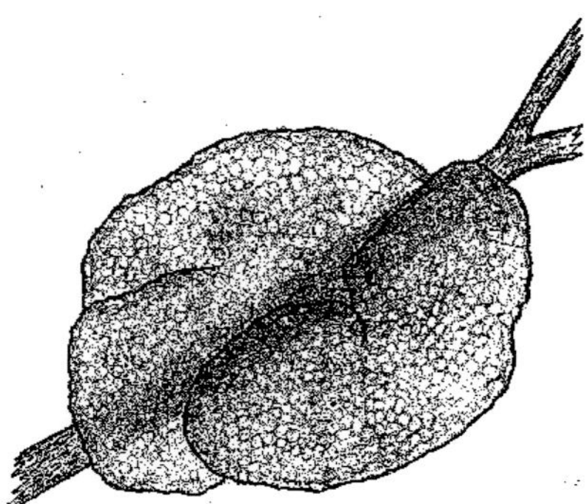

Bryozoan (20%)

Small Crustaceans

Several types of small crustaceans are caught with a seine and dip net. Isopods, scuds and fairy shrimps are common catches. Rare are tadpole shrimp, opossum shrimp and clam shrimps. Usually too small to notice are seed shrimps, copepods and water fleas. A small decapod, the grass shrimp, is also commonly caught while crawfishing. It and the larger crustaceans are discussed in the next section.

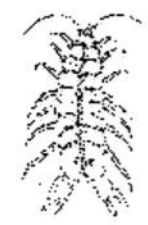

Isopods. Freshwater isopods are smaller than land (pill bugs) and marine isopods. The body shape is flat, from top to bottom. The typical color is brown. Isopods are very common in dip net catches. Adults are 0.5 inch long.

side view

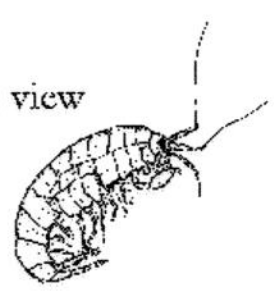

top view

Scuds (sideswimmers). These small crustaceans are fast swimmers. A scud's body is somewhat thickened from side to side. The color is usually tan. Adults are 0.5 inch long.

Fairy shrimps. Fairy shrimp do not have a carapace and adults are commonly 1 inch long. They glide through the water swimming legs-up. Their glistening bodies, lighter in contrast to the dark background of standing water or litter, are easy to spot.

Clam shrimps. The clam shrimps have a carapace that is compressed from the sides to cover most of their typical 0.5 inch long body. The carapace structure makes the animal look much like a clam. Their color may be bright red or greenish.

Tadpole and opossum shrimps. The tadpole shrimps have a carapace that looks much like that of the marine horseshoe crab. They are unlikely catches and, if they are seen it will be in the western part of the state. The adult length is commonly 1 inch. Opossum shrimps are small, slender crustaceans that look similar to marine decapod shrimps but without walking legs and with much larger eyes. They usually grow to 0.5 inch. Opossum shrimps live near the coastline and are unlikely crawfishing catches.

Large crustaceans

Larger freshwater crustaceans include many crawfish and a few species of shrimps. Certain crabs temporarily invade freshwater habitats along the coast. All are decapods.

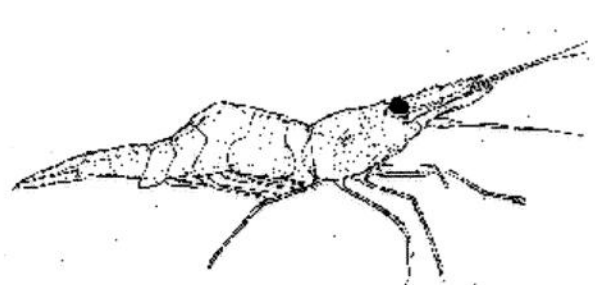

Grass shrimps. The small and translucent grass shrimps (also called glass shrimps) are hard to see except when they begin to jump in the net. Females that carry the green colored eggs on their swimmerets are more obvious. Adults are usually less than 1.5 inch long.

An abundance of grass shrimp is sometime seen in the shallow vegetation of streams, stream overflows and the ditches and shallow water shorelines of impounded streams. The most common one in the central United States and in Texas inland waters is *Palaemonetes kadiakensis*. Other, rarer species are found in and near openings of springs and caves.

River shrimps. River shrimps are found in stream basins from Virginia to South America that drain into the Gulf of Mexico and Atlantic Ocean. Like some fish, they migrate upstream. River shrimps grow form early juveniles to adult stage in the freshwater of river basins. Egg bearing females return to marine water to release hatchlings, which have a salty water requirement. The small shrimp soon migrate up freshwater streams to grow and complete the life cycle. Reservoir dams provide a difficult or impossible barrier for river shrimp to surmount. River shrimps are thus able to live only in the stretch of a river basin that lies below the first major reservoir. Males grow larger than females.

Four river shrimps live in U.S. rivers:
Macrobrachium ohione
Macrobrachium olfersii
Macrobrachium acanthurus
Macrobrachium carcinus

All four river shrimps live in Texas waters. All have no popular or common name. The first two species listed above are smaller than the last two. The first, *ohione*, has the widest distribution and is most commonly caught.

Claw (second leg pair) characteristics of adult males are usually sufficient to distinguish the four river shrimps. *M. olfersii* has a spiny large claw and a less ornate small one. The long slender claws of *M. acanthurus* have an obvious velvet pile on the pincers of larger specimens. The joint behind the pincher joint is about as long as the pincer joint (palm plus finger). *M. ohione* has bristles but no velvet pile on the claw. *M. carcinus* has more robust claws. Its second claw joint is shorter than the one consisting of palm and finger. *M. carcinus* is largest Texas species reaching a body length of 10 inches.

River shrimps have color cells in the skin that expand causing the whole animal to better blend with the background color of its habitat. They are darker, for example, in waters with black or brown bottoms or debris, and lighter in brighter surroundings. Younger animals are less adept at changing to a darker hue than older ones. Most young river shrimp appear translucent like the grass shrimp.

Both grass shrimp and river shrimp have distinctive characters: (1) a second leg pair that is larger than the first; (2) pincers present on the first two leg pairs; and (3) a second abdominal plate that overlaps, on the sides, the plates in front and behind of it. When the shrimp crawls, the first leg pair is typically folded inward under the head and the second pair is extended forward. Grass shrimps are smaller than river shrimps and the legs of the second pair are much more thin and delicate than those of river shrimps. Grass shrimps are much more common than river shrimps and their widespread distributions reach far upward into river basins.

Crabs. Certain marine crabs sometime venture long distances from saltwater via freshwater waterways. Blue crabs are notorious for this. A 5 inches wide blue crab in a dip net is a sure surprise to an inland crawfisher. Hermit crabs colonize shores of freshwater creeks and canals of the flat coastal plain far from saltwater.

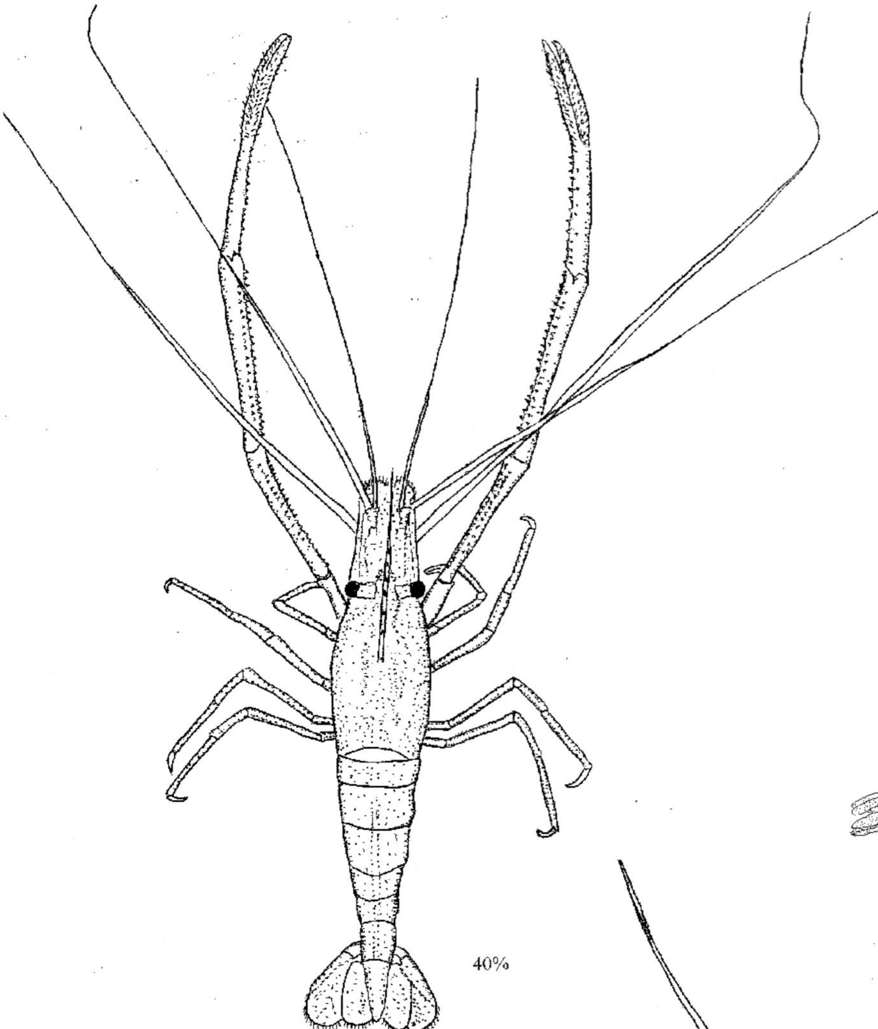

40%

Macrobrachium acanthurus male

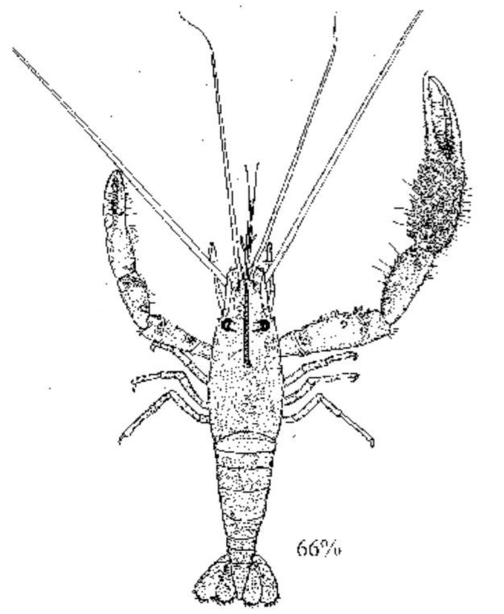

66%

Macrobrachium olfersii male

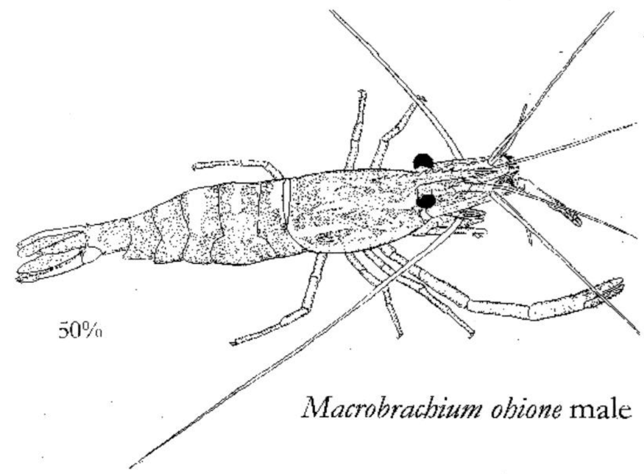

50%

Macrobrachium ohione male

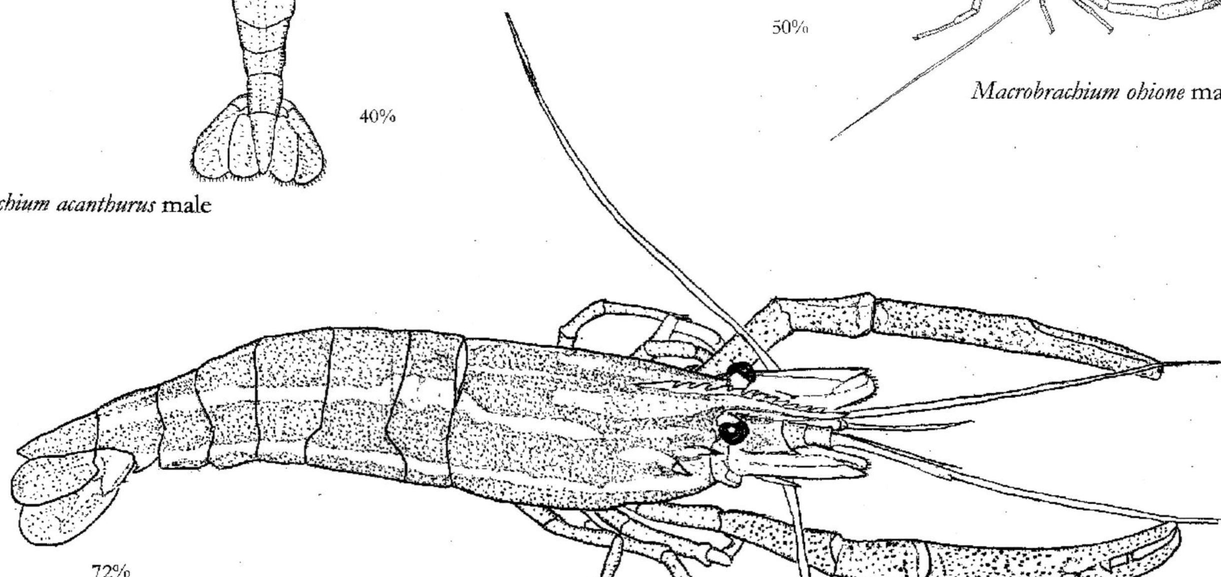

72%

Macrobrachium carcinus male

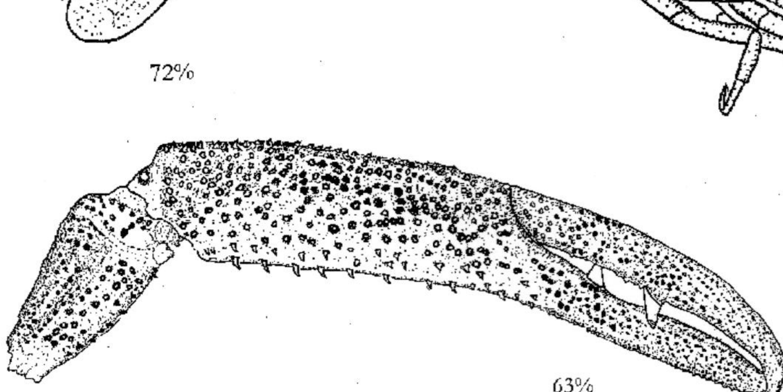

63%

Macrobrachium carcinus claw from large male

Live Collections

Keeping a group of crawfish alive as a collection has many benefits beyond just the enjoyment of their beauty and behavior. Important examples are:

1. The experience grounds one with a stronger familiarity of the varieties and a confidence in species identification.

2. Small specimens may be grown over several months until they reach preferred size.

3. The eventual molting of captives results in more perfect specimens. New shells are free from stains and precipitates that normally darken and distort the appearance of long-used shells. Also, the molting partially, or sometimes completely, restores lost or damaged appendages.

4. In the rush to catch and store live animals at the catching site, time may not be available to examine each male specimen to see if it is a Form I or II. Transfer to a live collection allows time for any Form II specimen to molt into a Form I animal, the stage necessary for proper identification.

5. Keeping specimens for a while in good conditions also allows for better photography. The photographer is less rushed and has the subject in a more controlled setting.

6. Those unfamiliar with keeping aquatic animals can learn to do so.

Maintaining specimens

Maintaining captive crawfish requires feeding and providing good water conditions. An excellent maintenance container for live specimens is a plastic shoe box. The lids snap shut to prevent escape but not so tight to prevent sufficient oxygen entry from the outside. Water is added to a level that is at or just above the top of the carapace. This allows the crawfish to expose gills to the surface should this become necessary. Water should be exchanged every two or three weeks. Chlorinated tap water is suitable if a tiny amount of waste is left in the container to combine with the chlorine at the time of exchange. Ornamental fish ponds and rain barrels are quick and easy supplies of good quality water. Ten to twenty containers of crawfish are easily maintained but fifty or more becomes a struggle.

A suitable diet is ornamental fish food consisting of pellets no larger than a BB, fresh spinach and carrot scrapings. Carrot scrapings and spinach pieces do not foul the water and neither do pellets if they are not over fed. Scrape carrots with a knife blade. A weekly feeding provides a suitable maintenance diet for crawfishes other than tiny ones that may need food every third day. A growth diet requires feeding at three or four day intervals and weekly water exchange. Ten days without food is too long to keep crawfish healthy. The quantity of food to be provided is one that would fill their guts about twice.

In a holding box, most crawfish will gradually consume their shed exoskeleton over the course of a couple of days. Should they be of some value, gonopods may be removed before they are eaten along with the rest of the shed exoskeleton.

The containers of live animals should be marked properly to prevent confusion. The information should be written in permanent ink so that it does not wash off. Writing data on the container body instead of the lid prevents errors that occur when lids are accidentally switched. Permanent ink markings that accumulate on reused containers may be removed easily with all-purpose spot remover or suitably with ethyl alcohol.

The live collection may be kept indoors or outdoors. If outdoors, the shoe boxes or other vented containers must be protected from predators such as birds, raccoons and cats. We place the crawfish boxes within large plastic storage boxes that snap shut and also allow sufficient entry of air around the lid. These are placed in a spot that provides continual shade. In these conditions, crawfish do well even in daytime temperatures that reach 100 degrees Fahrenheit. In winter, they are brought indoors before freezing weather arrives.

Above: Live collection set up. Shoe boxes with crawfish therein are in the storage boxes under the shade cover.

Below: Ornamental pond. A good water supply for a collection of live crawfish.

Keeping Crawdads for Display

Crawfish are displayed for the same reasons people display ornamental fishes and other small animals. They provide live subjects for a hobby. The effort of keeping them alive and showing them brings educational and aesthetic values. Their beauty may be enjoyed for an extended period in a domestic setting.

Crawfish may be displayed in an aquarium in the usual manner suitable for fish. Units must be covered completely or the crawfish will crawl up air hoses or other structures and escape. Fish food composed of small pellets, spinach pieces or carrot scrapings may be given to hungry crawdads.

A large bowl with edges high enough to prevent escape is also useful as a display container. In bowls, crawfish may be observed from the top and side, which are the best viewing angles. A cover may be used to reduce evaporation and protect the crawdads from cats, dogs and small children. Water level in the bowl should be kept just above the top of the shell of a resting animal. If water fouling is a concern, the animals may be removed to a separate container for feeding or bowl cleaning and then returned to the display unit.

Hobby Growing

Crawfish will thrive in ornamental fish ponds with hard bottom liners. Covers are needed to protect them from predators and prevent their escape. They will eat valuable plant specimens and consume fish eggs and small fish. Otherwise, they are generally harmless. Crawfish grow better when kept in a natural pond setting. Being bottom crawlers, the depth, however, is often too great for visual enjoyment.

Crawfish can be kept and grown in plastic (hard) swimming pools. Suitable structures should be given for hiding places. A hose may be used periodically to wash out the units. Chlorinated water is fine if some residual of the organic material is present at the time of exchange. Most important is that outside units will be vandalized and the crawfish eaten by predators, particularly raccoons, if they are without very sturdy covers.

Crawfish may also be grown in smaller units, as mentioned, for live collections. Regular water exchange and feeding are necessary. Food selection is also important in small units. Carrots, spinach and some pelleted foods are preferable because they retain their integrity until eaten. Some burrowers may be slow to accept carrots, spinach and pellets.

Crawdads as Fish Bait

Crawfish are excellent bait for sport fishes. Most bait shops carry them and sell them by count or by weight. The most available species is the red crawfish. Fishermen often use crawfish to catch catfish and bass. For bass and catfish, an adult crawfish is hooked by inserting the sharp point of the hook about midway beneath the tail and directing it upward until it emerges through the hard shell on the tail's top. The bait is then cast to a good spot. A float may be used to suspend the crawfish off of the bottom.

Crawfish are also a favorite bait for catfish when trotline fishing. A trotline is a long line of 25 to 100 feet in length with short hook lines affixed every 2 to 3 feet. A trotline is set by attaching the ends to floats or to trees so that the main line is suspended near the water surface and is in a straight line between the attachments. To fish the line, the short hook lines are baited with juvenile or adult crawfish. The bait is attached as mentioned above or, in the case of smaller juveniles, threaded end to end as one would hook a worm. A trotline is normally fished from a boat. The boat is pulled along parallel with the path of the long line while removing and boating any hooked fish and re-baiting empty hooks.

Small juvenile crawfish (and grass shrimp) are used with pole and line to catch sunfish such as bluegills and crappie. The shape, color and chemical attraction makes the crawfish irresistible to this kind of fish. Fishermen not familiar with the use of this bait will be amazed to see how fast they can catch nearby fish. Sizable numbers of small bait crawfish may be caught in winter and spring from shallow ditches.

Photographing Crawdads

Making photos of crawfish is a great way to have permanent enjoyment of the beauty of crawfish. A set of reference images provides many of the same benefits of an alcohol-fixed specimen collection. Additionally, photographs give a lasting record of color.

Photography

Good crawfish photos can be made with many of the film and digital cameras available on today's market. Crawfish photos of exceptional quality are possible if cameras have the required features. Proper equipment, however, must be matched with good technique to create such images. After some trial and error, the details are worked out for shooting photos with a particular camera setup. A good crawfish photo has several desirable qualities:

The specimen is the primary subject of the photo.
The image is very sharp (not fuzzy or blurry).
The whole animal is in focus.
The image has true color and good contrast.
The specimen has little or no surface glare

Primary subject

The photo should mainly show the animal. In photography language, the animal should fill the frame. Surroundings are thereby kept to the minimum.

The nearest focus point of many cameras is too far away from small subjects like crawfish to produce an image that has an in-focus subject filling the frame. This means that the closest focus point yields only a small crawfish image in the middle of the picture. Some adjustment can be made by cropping, but when the image is enlarged for re-size, some loss in quality occurs. Certain cameras have a zoom feature that may be used to increase the subject's portion in the frame. The closest focus point of many zoom cameras, however, may still be many inches from the lens. Check the focus distance when selecting the camera.

Cameras with a through-the-lens focus produce a photo true to the view. If the viewfinder does not show a through-the-lens image, a close-up shot has a slight aberration: the view that fills the viewfinder will not be the one that shows in the photo. The photo's subject is somewhat off center and maybe partially out of view. This is because the aim of the viewfinder is different from that of the lens. When selecting cameras, consider how distance affects both focus and view.

Sharp images

Digital cameras capture the image on a sensor. Film cameras capture the image on film. The snap happens when the camera opens to let light in through the lens. It is the moment of image capture. The entering light is the light reflected from the subject and its background. Two things may be in motion at the moment of snap which may blur the captured image: the camera and the subject. A fast snap decreases the influence of movements.

The photographer holds a camera steady. Photographer movement is much less important at a snap speed of 1/250 second than one second. For hand-held snaps, a 1/60 second snap speed is usually good enough if a flash is used to make film shots of live crawfish at rest. Tripods give complete steadiness and permit the use of slower snap speeds.

The capturing medium (film or sensor) has some influence on the sharpness of the photo. Digital cameras have different degrees of resolution capability. Camera cost usually rises with a greater degree of resolution. The film largely determines the resolution capability of film cameras. Slow films produce greater sharpness than fast films. Speed ratings of commonly available roll film range from 25 to 1000 ASA. Higher numbers mean faster film — film requiring less light for proper exposure. A film speed of 400 ASA is usually the upper limit for very sharp print images (4" X 6"). As an image is enlarged, its prints lose sharpness. When reduced in size, images usually gain noticeable sharpness.

Making photos of freshly killed crawfish in posed positions can eliminate blurring due to subject movement. A film camera is mounted on a tripod and the photo snapped at slow exposure speed using slow film. Many of the crawfish images in print today were made this way.

Live crawfish were used to make images for this book. Suitable photos of live animals are made with patience. Certain specimens are difficult subjects. Placing them directly into chilled water produces temporary immobilization.

Less expensive digital cameras are not suitable for imaging live crawfish. The delay that occurs between button press and the snap gives a momentarily still subject time to regain movement and change position.

Focus of a whole animal

A good depth of field is required to capture the entire animal in focus. Depth of field is the distance range between the nearest and farthest objects in a scene that are of an acceptable focus. Imagine, for example, if one uses two cameras to focus on five marbles lined up end to end and away from the photographer. One camera produces an image in which only the center marble is in a clear focus. The second camera produces an image in which the second to fourth marbles are in clear focus. The second camera has a better depth of field. Good crawfish photos are made with cameras having sufficient depth of field to show the clear image of the creature front to back and top to bottom.

Given that a camera has the potential for making such an image, it is helpful to know how to use it for best results. The lens opening of a manually adjustable camera can be set to gain a greater depth of field. The depth of field is increased by decreasing the lens opening (that is, the opening produced when the camera snaps). Decreasing the opening, however, means that less light will pass through the lens to be captured by the camera.

To make up for less light, a couple of things can be tried. If film is used, select a fast film that promises high definition. The brightness of the subject can also be increased by use of electronic flash. Most of the color photos in this book were made with 400 ASA film, 1/60 second exposure, f/32 lens opening and a single electronic flash.

Depth of field is also greater when the subject to camera distance is greater. For example, when in focus, the subject snapped at a distance of 2 feet has a greater depth of field than one snapped at 6 inches from the camera lens. Macro (not close-up) lenses fill the frame with a crawfish subject while still remaining more than 1 foot from the animal.

Another determinant of depth of field is the length of space between the lens and the capture medium that the exposing light rays must pass within the camera. This distance, called the focal length, is either fixed, as in fixed lens cameras or variable in cameras with a movable lens, such as a zoom lens. Depth of field is greatest at short focal lengths. The focal length is not normally an adjustable feature of a particular camera.

Close up lenses are sold for attachment to the outer face of a camera lens. These are essentially magnifiers that cause a loss to depth of field as the magnification increases.

A large crawfish is about 1.25 inch tall at rest. The height of small crawfish is about 0.25 inch and medium sizes are 0.5 to 0.75 inch. Widths are about two to three times the heights.

True color and good contrast

Properly used, film and digital cameras give images that are true to the color and shade characteristics of the animal. Choice of background is important. A black background produces good patterns and colors even in very dark animals. A light (not white) background helps to define margins in small specimens that are partially transparent.

Properly exposed film (not over or under exposed) provides a good negative. Present day, neighborhood film processors all use chemicals to develop the film and then scan the negatives to produce a digital form before printing the photographs. Color adjustments, good or poor, occur after the scanning step. It is usually the competency and attention of personnel, not the equipment, that determines the quality of the final product.

No glare

Light easily reflects from the surface of crawfish causing a glare that shows up in photos. Photographing crawfish subjects while they are covered with water eliminates the problem of body glare. The animal is submerged in shallow water in a small container such as a plastic shoe box. For background, we use a black felt cloth placed under the crawfish at the water bottom.

Light reflection from the water surface shows up in the photo if proper measures are not taken. Position the container so that sky light does not enter directly from above. A set up under a porch cover or just inside a window is usually sufficient. Covers may be used outdoors. Some outdoor light is needed to see the animal through the camera's viewfinder. When using an electronic flash, aim the direction of the flash so that the burst of light hits the water surface at an angle sending the reflection of the flash apparatus away, and not directly into, the camera lens.

Recording photo information

Confusion regarding the what, when, where and how of crawfish photos is avoided by taking notes as the photos are shot. Here is a list of items to record as a set of crawfish images is made.

- Date the roll is begun
- Location and date of capture for each subject
- Species and sex of each
- Total length of each
- Number of frames taken per animal
- Notation about any unusual specimen (i.e., a small left claw)

Soon after the prints are made, write the following information on the back of each print.

- Location and date of specimen capture
- Species, total length and sex
- In the case of film, development number (number attached to negative roll)

Information about each photo set may also be entered into a record book or file. This record is useful for associating images with specimens in permanent (preserved) collections. An image record helps to direct the collector to the exact specimen, the species identity of which may not have been known or confirmed when the photo was made.

Drawing Crawdads

Drawings have many of the same benefits as photos. In addition, drawings may better illustrate color and contrast of some species, and the detail of small specimens. A certain satisfaction is also obtained from the creation of a good crawfish drawing.

Color drawings

Drawing is done in many ways using many kinds of materials. Skill and talent is required to make sketches and freehand a drawing into its final form. Here, we are not writing about how to achieve the level of drawing skill that takes years to develop. Rather, we propose shortcuts to creating decent color images by the use of tracing, low-cost materials and a copy machine. The goal is to make pencil drawings on paper that show crawfish images which are realistic and accurate in proportion and color. This is more like drafting than drawing in the arty sense. Good results are achieved almost immediately. If one wants to learn more about the many techniques and variations of pencil drawing, many good books are available.

Materials

Materials are few: photograph, regular lead pencil, box of colored pencils, sharpener, eraser, masking tape and semi-transparent paper.

Procedure

Select a photo of the crawfish you want to draw. Use a copy machine to make an enlarged, black and white copy of the photo. Copy machines with a special feature for photographs work best. The enlargement should be of a manageable size. A typical enlargement will increase the crawfish so that it fills a sheet of 8.5 X 11 inches.

A trace drawing is next made of the photocopy image. Tape a sheet of semi-transparent drawing paper over the copy image. A good paper is 100% cotton rag tracing vellum. Use a soft lead pencil to trace the edges of the animal's shell and interior details such as color pattern, joints and obvious shell delineations. Use little pressure to produce thin lines. Fringe setae can be entered last using a very sharp pencil. The length of fringe setae should be marked during the tracing step.

We suggest that the final color drawing be produced directly on this tracing. The thin paper can be backed with white matting to disguise its transparency. If a color copy is made as a print after the drawing is finished, a completely different paper will result anyway.

An initial tracing can also be transferred to sturdy art paper. One way to do this is to trace with a graphite pencil on vellum and then, after flipping and then taping the tracing to the new paper, trace again over the backside to transfer the graphite lines. Again, we suggest that beginners use the initial trace without transfer to develop the final color drawing. Later, they may want to use the transfer technique.

Continuing then with the tracing on semi-transparent papers: Recreate the colors of the animal with color pencils. The color photo again becomes useful at this point. Begin by testing mixes of pencil markings on a test sheet until color matches are found that are identical to those in the photo. Crawfish usually have a base color that covers most of the shell. Try to duplicate that color and determine which colors to layer or not to layer over it to get the desired end color. Several layers will be required to replicate dark or more intense colors. Determine all colors before beginning to color the drawing. Add the base color and then the layering needed to color the more prominent shell markings, joints, tips and segments. The tracing marks are gently erased as the drawing is made.

Consider darkening the margins of transparent parts such as the appendages. Margins look somewhat darker in the three-dimensional view, the view we see with our eyes. This is because the curving edges have less translucence and contrast sharply with the surroundings. Light reflection also contributes to darker edges if light is supplied from the direction of the viewer. If the background surrounding the crawfish subject is dark, scratching some color off of the image at its edge may produce a better light to dark contrast.

Colored pencils are generally of two types: oily or chalky. The oily pencils produce deeper colors but are not as easy to erase. Most advanced pencil artists use the oily pencils and apply several techniques to blend the colors by dissolving them. We think that beginners do better with the chalky pencils. Errors can be fixed easily with an eraser and satisfactory colors result from careful use of layering. Chalky types are bought at stores having school or hobby supplies and are inexpensive. Colored pencils need to be sharp so a good sharpener should be available. A block or pencil-style eraser is also handy.

The drawing is complete when a final touch-up conforms the color of the drawing's specimen to that of the original photo. A copy of the drawing may then be made with a color copier at the local copy center. When making the color copy, the drawing can be reduced to a more desirable size. By reducing the size, imperfections in the drawing are disguised and grainy color is lessened or lost. This is one reason that the photo was enlarged in the first step of the process.

Add data about collection locality, species, sex, length and location of the photograph to the back of the drawing and its copies. Any one can then associate the drawing to its history and the subject to its full identity.

Computer Applications

Computer applications are available for producing images both at home or professionally. Technology is advancing so rapidly that one can hardly keep up. Applications are available to make drawings directly from photographs. These, and a number of other software applications may be used to create an array of enhancements for images. Even depth of field of digital images may be increased. Those with access to various equipment and software may find enjoyment in using them to produce a collection of enhanced prints or drawings. Some devices, such as do-it-yourself photo scanners and printers, are available at most photo processing outlets.

Reference Collections

A collection of chemically fixed crawfish is a specimen source useful for comparison to future catches. Usefulness is dependent on proper identity of specimens and valid collection data. Properly set up, a reference collection requires little maintenance and is long lasting.

Specimen preservation

Biologists generally use a 5 - 6% formalin solution to initially kill and fix the specimens. After a few days, this liquid is discarded and the specimens are washed for a few hours in running tap water. Finally the specimens are submerged in 70% ethyl alcohol and the container is sealed. Nonprofessionals usually do not have formalin readily available for purchase. Rubbing alcohol in the form of 70% ethyl or isopropyl alcohol is available from most pharmacies. Ethyl costs a little more than isopropyl but is the better fixative.

We have tried many ways to accomplish humane killing of specimens and have found the following method most suitable. The container that is to receive the specimen is mostly filled with alcohol and placed in a freezer or freezer compartment of a refrigerator. An hour (or more) later, another container containing the live specimen and enough water to just cover it is placed in the freezer also. After some time, the specimen container is checked for the presence of frozen water. When a sheet of ice has formed around the edges and above the specimen, the specimen is removed and placed into the chilled alcohol. This is kept out of the freezer or placed in the refrigerator to retain the chill. Not forgetting specimens in the freezer avoids a hard freeze of specimens, subsequent brittleness and loss of appendages.

Specimen storage

Excellent specimens are best kept individually. Container type is very important. Jars need to have wide mouths and be of a size that will contain enough fixative to do its job, but not be too large so that fixative is wasted. The containers must have airtight lids. Pint size canning jars and smaller make good containers for most crawfish.

Organizing a collection of fixed specimens takes effort but has a big pay off in convenience and the avoidance of confusion and errors. Four kinds of information are written for each specimen. One is an entry in the field log that is taken along on the crawfishing trip. The field log records data about the place where animals are caught, date caught, how caught, what was kept and perhaps some notes on the habitat. The second organizational information is written as a mark on the field container. It indicates the place of capture and helps to avoid mix ups. The third organizational information is the entry into a collection log. The collection log is a record for all specimens contained in the reference collection. This entry records all pertinent information about a retained specimen. The fourth organizational information is written on and in the permanent container of the new specimen.

The collection log may be set up with written columns. The first column is an identifier such as the collection number. The second column gives a general location such as a county. The third column includes an array of data for the specimen that includes when and where it was caught, the collectors and means of collection, the sex, the size and the stage (Form I or II, juvenile or adult). The next column gives the species identification. A last column can give extra information such as photo numbers or habitat features.

The container that holds fixed specimens in a collection should have data on the inside and outside. A collection number is sufficient for the outside designation. This number can have an associated code letter that assigns the specimen to a group, such as P for *Procambarus*. The written information placed inside the container and in the fixative with the specimen is the most important. It has the same number as on the outside plus the following pieces of information: species, collection locality, date of collection and collectors. Miscellaneous information can be written on the back of this collection card. Cards should be cut from good quality paper and entries made with permanent, water proof, fade proof ink.

Specimen examination

Several pieces of equipment aid the examination of fixed specimens: a pan of suitable size, forceps, a probe, a measuring device, a roll of paper towels, a table lamp and a magnifier. Hand-held magnifiers of 15X are best but 10X will do -- Even better is a dissecting microscope. A probe can be any pointed device that is useful to gently pry a gonopod away from its resting place so that it can be observed in side view. The probe does not remove the gonopod. If a gonopod is removed with forceps for examination, some method should be devised to ensure that the now loose appendage is identifiable in the future as a part of the exact specimen from which it was removed. Some people place detached gonopods in numbered vials and then back into the specimen container. Others mount them on insect pins with a descriptive tag that clearly associates the gonopod with its specimen. To prepare a gonopod for pin mounting, it is first placed in alcohol to displace its internal water and then into xylene to clear the tissue.

Some members of the genus *Procambarus* (slenderclaws and southern streamers) bear a tuft of brush-like setae near the terminus of the gonopod. This tuft partially or completely obscures the terminal elements. Removal of the setae must be done under magnification while holding the gonopod with forceps. A fine needle is used to carefully tear the setae away without damaging the terminal elements.

Commercial Production

Food aquaculture

Most U.S. crawfish production for food occurs in south Louisiana and southeast Texas within 100 miles of the Gulf of Mexico. The land in this area is flat and surface water is usually plentiful. Growing ponds can be quickly filled with freshwater available from canals that inundate the region. The climate is suitable for rapid growth of crawfish. Clayey soils and a water table that is less than 3 feet below the land surface are other favorable features of most areas of the region.

Low levees are built to contain water 1.5 to 2 feet deep. The levee stands about 1 foot above the water surface. Some crawfish ponds are built around low areas or backwaters and are often a combination of natural and artificial creations. Though some levees are built around wooded areas, most are around open, treeless areas. The open ponds are better producers. Pumping surface water from a nearby source fills the pond and maintains depth.

Pond culture usually takes advantage of the burrowing behavior of crawfish and mimics the seasonal dry-wet cycle common to the southern Unite States. Crawfish are grown while water is present in the pond from autumn to spring. As the warm, dry season approaches, ponds are slowly drained (over two to four weeks) to encourage the burrowing of adult crawfish. The burrowing population consists of enough brood crawfish to provide a new hatch of young when spring re-flooding again provides standing water to the ponds. The burrowing population contains residual, uncropped crawfish of the previous cycle, plus new adults from neighboring ponds that were added at the beginning of the draw down.

Crawfish feed on vegetation, small animals and decomposing organic matter. In pond culture, forage is artificially provided for the crawfish crop. Many types of vegetation are suitable, but rice is the preferred plant. Planting and growing of forage occurs while the ponds are dry. When the pond is flooded in autumn, emerging young eat the submerged forage and the growing period begins.

When the young reach harvestable size, they are trapped. The first young are ready for catching in two to four months. Subsequent hatches during late fall and mid winter may require four to five months to reach harvestable size. The trapping commonly begins in late September or early November and continues until May.

Trapping takes place mostly from fall to spring. Traps, cylindrical or pyramidal in shape, have small inverted cone entries that provide underwater access by crawfish. A trap's open end rises above the water surface. The closed end sits on the bottom. The walls have a mesh size that retains larger crawfish and bait. A boat is normally used in trap retrieval. Traps are set in a line. As the boat passes, the traps are lifted, catch dumped, new bait is added and the traps are returned to the pond in one synchronized motion.

Catch is transferred to submerged purging pens where the crawfish's intestines empty within 12 hours. The crawfish are washed and placed alive into shipping containers, usually porous sacks (onion bags).

Open ponds are sometime used to double crop crawfish and rice. Normally, rice is planted in a dry field in early spring. When it reaches several inches, the field is flooded. Near harvest time, the fields are drained and left to dry for a couple of weeks. Harvest is done on dry land with a combine. The time span for planting to harvest is only three to four months.

In rice crawfish cropping, crawfish adults are stocked when the young rice is flooded. These crawfish later burrow when the ponds are drained for rice harvest. After harvest of the rice, the ponds are again flooded. The rice stubble and other organic growth provide forage for the crawfish crop. Higher marginal levees provide a deeper flooding for crawfish than internal rice levees. Crawfish are trapped through winter and the cycle begins again in early spring. Some rice varieties are chosen for their suitability as crawfish forage.

The red crawfish (*Procambarus clarkii*) is the primary species of culture and *Procambarus zonangulus* is the secondary species.

Crawfish culture has complexities and details that the above account does not address. Information about water management schedules, stocking, forage selection, harvest techniques, marketing and many other methodologies can be obtained from county extension offices and crawfish farming events.
In recent years, China has become a large producer of crawfish (the American

red crawfish). Much of the tail meat sold in the U.S. now comes from that country. If a cooked product in the form of whole crawfish is seen in markets, it is usually from a Chinese source.

Wild Catch

Wild harvest provides a substantial portion of the food crawfish market. It varies from year to year according to climatic conditions. The Atchafalaya River basin in south central Louisiana is the major producer of wild caught crawfish. Crawfish (primarily red crawfish) are trapped in abundance when basin flooding cycles support great numbers of large crawfish. Wild crawfish are mostly caught with baited pillow traps in shallow water. These traps submerge completely and have funnel inlets in two of the corners. The catch is marketed in sacks like the farm-raised crawfish.

Other products

Live crawfish for fishing bait are supplied by both wild and cultured sources. In aquaculture, the surplus of smaller animals is often made available to this market. The red crawfish is the primary bait product in the south. In the north-central United States, *Orconectes virilis* is frequently found in bait shops.

Aquaculture provides a limited number of specimens for aquarium shops. These are usually bright blue crawfish from Florida (*Procambarus alleni*). A more limited product is the science specimen. Specimens are sold to educational institutions in chemically preserved lots.

Byproducts of processing plants produced from shells, heads and small animals are of some commercial value. Chitosan is used in the food industry and astaxantin as an antioxidant for many purposes. Some byproduct is processed for use in animal feed.

Cooking Crawfish

Americans eat the fleshy or muscular part of crawfish. This usable tissue is found in the tail. The tail meat makes up 10 to 20% of the body weight, usually 10% in larger crawfish. That means one needs to peel 10 pounds of tail of larger specimens to get 1 pound of meat. The tail meat of marine shrimp, by comparison, contributes about 50-60% of the whole animal weight.

Crawfish are available in the market mostly as packaged frozen meat or live whole animals. Sometime they can be bought in a precooked style, usually as whole boiled.

Restaurants commonly prepare dishes that contain crawfish according to recipes that are also used for shrimp. Post molt blue crabs are sold as a delicacy using the name "soft-shell crab." Such crabs are prepared as a pan-fried dish. Crawfish are also sold for this purpose.

Two preparations traditionally identified with crawfish are etouffee (pronounced ay-too-fay) and boiled crawfish. Etouffee, a dish common in home cooking, is prepared with tail meat. The whole live animal is the basic ingredient of a "crawfish boil." Family picnics and large gatherings are events that often include a crawfish boil as the main attraction. Preparation methods for these traditional ways of eating crawfish are given below.

Other crawfish dishes popularized by folk song and chefs are not as common for home fare. Examples are crawfish cornbread, crawfish po boy, crawfish pie, crawfish gumbo, crawfish bisque and crawfish jambalaya. Jambalaya and gumbo are more often made with shrimp and combined with cooked rice. Bisque, a thick soup, is prepared by making a stock from tail meat and shells. The stock is combined in various ways with things like oyster liquor, a roux, mashed rice, onions, celery, garlic, parsley, butter or vegetable oil and various seasonings.

The best way to have a crawfish boil is to buy a suitable amount of live, farm-raised crawfish at the market and prepare them outdoors over a propane cooker using the method given below or some variation of it.

It is unlikely that an adequate quantity of crawfish will be caught using most of the methods found in this book. If one aims to boil up less than a suitable amount and to do so over a wood fire in the backyard, use clean water, a pot that will not cause metal poisoning, some spice and a cooking time able to kill everything living on and in the crawfishes.

Boiled crawfish

Equipment:

- Large boiling pot of four to five gallon size and lid
- Outdoor propane cooker
- Wire basket insert for pot
- A styrene ice chest (high heat warps plastic chests)
- Towels or pads to grip pot lid and basket

Ingredients:

- 10 to 20 pounds of whole live crawfish, thoroughly washed in fresh water
- 2 or 3 gallons of water – pot half full
- Crawfish/crab boil seasoning – liquid or dry, or both
- Eight onions cut in half
- Two or three lemons cut in half
- Two stalks of celery cut up
- Five ears of corn each broken into two smaller pieces
- Eight small potatoes (new potatoes) washed and skin left on

Preparation:

1. Heat the water to boiling
2. Add seasoning according to package instructions
3. Squeeze lemons to add juice to seasoned water
4. Boil for five minutes
5. Add vegetables and lemon halves to basket
6. Boil for five minutes
7. Remove and place in ice chest with lid open, cover with paper
8. Turn up heat
9. Add crawfish to basket and place in pot
10. Boil (rapid) for five minutes
11. Kill boil with 4 pounds of ice
12. Add previously cooked vegetables to crawfish and let soak with crawfish for 20 to 30 minutes, stirring occasionally to redistribute heat
13. Remove from liquid and serve

Serving:

Dump crawfish into piles on a table covered with thick paper or old newspaper – paper can be rolled up at clean up time
Though already seasoned, some people like cocktail sauce with their crawfish

Eating:

Tail meat is easily removed from the shell as follows:

1. Hold the head (the carapace) on the sides and from the front with one hand.
2. Grab the tail with the thumb on the underside.
3. Gently twist the tail away from you until it breaks free from the head.
4. Peel away the first two shell segments of the tail to expose the meat.
5. Pinch the tail in front of the fan and push the meat out with the thumb or
6. Pinch and then remove the meat by pulling the exposed meat portion
7. Consume the meat directly or after dipping in sauce.

Etouffee

There are many variations of this recipe. The traditional ones include the making of a roux as one would do when preparing an authentic gumbo. One would expect this type of dish in a restaurant. Quicker home preparations use mushroom soup instead of a roux. Also quick, is to merely saute the vegetables, add some canned tomatoes containing peppers, cook down the mixture and add the crawfish meat. Etouffee is usually served over cooked rice. The following is an example of one of the home style quick versions.

Equipment and Ingredients:

A deep sided pot large enough to cook the desired number of servings.
2 pounds crawfish tail meat
1 cup chopped green onions and tops
1 cup chopped celery
½ cup chopped bell pepper
½ cup chopped fresh parsley
½ cup cooking oil
One can (8 oz.) Ro-tel tomatoes
One can (8 oz.) Tomato sauce
1 tbsp salt
1 ½ cups of water (or fish stock)

Preparation:

1. Saute vegetables in oil until tender.
2. Add remaining ingredients except crawfish
3. With occasional stirring, simmer until most of the water in the tomato juices cooks out
4. Add water and crawfish tail meat
5. Return to simmer and cook over medium heat for 10 to 15 minutes

Serving:

Serve in bowl over cooked rice
Makes 6 to 8 servings.

Texas Crawdads: Scientific Studies

A historical synopsis of the descriptive work concerning Texas crawfishes is now given. It is meant to clarify the present understanding of species distinctions and to reveal areas of lingering confusion. This section is written to the advanced student.

Historical overview

The earliest reports on Texas crawfish dealt with their discovery by English-speaking naturalists. Charles Girard, in 1852, wrote the type description of the red crawfish, *Procambarus clarkii,* using specimens of crawfish that were collected somewhere between San Antonio and El Paso. In 1870, Hermann Hagen produced a report that listed three species from the state. Scientists of the 20th century would accept only one of these, the red crawfish, as valid.

Walter Faxon, in 1885 listed again Hagen's three records and added records for what are now called *Procambarus simulans* and *Procambarus acutus*. In 1898, Faxon described a new species and a subspecies which were later known as *Orconectes difficilis* and *Orconectes palmeri longimanus* using material collected in the Red River in north Texas. In the same work, Faxon recorded what is now known as *Orconectes neglectus* using material collected at the same locality. The latter record was considered questionable by 20th century scientists. So, by the end of the 19th century eight species had been listed from Texas, five of which would be clearly established as Texas residents in the following century.

The list of crawfishes present in Texas was greatly advanced in the 20th century. Besides a number of new species from the state that would later be described, Texas records for species described elsewhere would be reported by biologists. Of particular usefulness, was a statewide review by George Penn and Horton Hobbs in 1958, a river study by Rollin Reimer and William Clark (1974, Navasota basin) and one by Horton Hobbs Jr. (1990, Neches basin) and several works by Albaugh, Reimer and Hobbs.

Penn and Hobbs (1958) listed eighteen Texas crawfishes. By studying new and increased numbers of specimens, biologists of the latter half of the 20th century would correct or reject misidentifications given to crawfish in some reports. Access to increased specimen number by specialists also led to recognition of new species within the arrays of similar specimens, specimens that otherwise would have continued to be assigned to an established species. Careful study also led to the realization that certain species had incorrectly accumulated more than one name. Most helpful to students during this period were two checklists prepared by Horton Hobbs (1974, 1989). Almost fifty years after the 1958 review, we list 37 Texas species.

Progress in the understanding of Texas' species groups

Cambarellus

The status of pygmy crawfishes in Texas is well established. In 1973, Albaugh and Black named the last of four Texas species and clarified pygmy distribution in the southwestern portion of the group's U. S. range. Faxon (1884) described the first U. S. *Cambarellus* species, *shufeldtii*, using specimens collected by others in the New Orleans area. The other two species from Texas, *puer* and *ninae*, were described by Hobbs (1945, 1950).

Cambarus

In our species review, we present, mostly for convenience, the devil crawfish as *Cambarus diogenes*. Two recognizable forms, which overlap in distribution, are present in the devil crawfish's range. Some students prefer to consider each a separate species. Others continue to recognize two varieties of the same species.

Girard first described the devil crawfish, *Cambarus diogenes,* from Washington, D.C. in 1852. In the century that followed, it was found to be distributed over much of the central part of the continent. In 1884, Faxon described a new variety (*ludoviciana*) from New Orleans. Until the latter part of the 20th century, two subspecies were recognized: *Cambarus diogenes diogenes* and *Cambarus diogenes ludovicianus*. Until recently, the subspecies *ludovicianus* was considered a variety local only to the New Orleans area. Now its distribution is known to be much wider in the Gulf Coastal Plain. Hobbs listed two separate species, *Cambarus diogenes* and *Cambarus ludovicianus,* in his 1989 checklist of American crawfishes.

Most specimens that have been collected in Texas are of the *diogenes ludovicianus* variety. It is obviously more colorful than the *diogenes diogenes* variety. We used a photograph of a male specimen of the less colorful variety from eastern Texas to make the drawing shown in the fatclaw section. The exact capture locality is unknown.

Fallicambarus

Hobbs (1969) established the genus *Fallicambarus* to include some species removed from the genus *Cambarus*. Texas members are *foidens*, *macneesei* and *devastator*. The oldest known of the three, *Fallicambarus foidens*, was first described by T. J. Cottle from Canada in 1863 as *Astacus foidens*. In subsequent years, a number of species names were recognized as synonyms of *foidens*. Specimens were reported over many years from coastal Texas and Louisiana as *Cambarus hedgpethi* Hobbs, 1948 and later as *Fallicambarus hedgpethi* (Hobbs) 1948. Hobbs and Robison determined in 1989 that the correct name for those specimens should be *Fallicambarus foidens*.

The two other species, *Fallicambarus macneesei* (Black) 1967 and *Fallicambarus devastator* Hobbs and Whiteman, 1987 were described from southwest Louisiana and eastern Texas respectively. Reimer, Johnson and Marcello (1975) extended the range of *macneesei* to Galveston County, Texas. Hobbs (1989) listed Brazoria County, but we think that he probably did so in error , meaning Galveston County instead. We found *macneesei* in Fort Bend County. It is remarkable that *devastator* did not receive a scientific description until 1987, considering its obvious presence in urban areas, unique appearance and potential as a pest on farm and ranch. In an interesting 1991 paper, Hobbs and Whiteman recorded many observations about this unique crawfish.

Faxonella

Faxonella clypeatus, named by W. P. Hay in 1899, is the oldest member of this group. His description was of a female caught seven years earlier at Bay St. Louis, Mississippi by the U. S. Biological Survey. In subsequent years, *clypeatus* was found to occur broadly throughout the southern States from South Carolina to Texas. Penn and Hobbs (1958) reported the first Texas record of *clypeatus* from Marion County. Rollin Reimer, in 1972, added additional records of *clypeatus* from the Sulphur River basin (Cass County).

Faxonella beyeri (Penn) 1950 was described from the Red River basin in northwestern Louisiana. Reimer (1972) reported *beyeri* from the Sabine River basin in Texas. Localities ranged from Harrison to Newton Counties.

Faxonella blairi Hayes and Reimer, 1977 was described using specimens from the Little River basin in southeastern Oklahoma (McCurtain County). In that description, an additional record located the species below the confluence of the Little River and the Red River in Arkansas. The new locality reported herein from Red River County is no more than 20 miles away from the one in Oklahoma. *Faxonella blairi* is very close in appearance to *Faxonella clypeatus* and may be a subspecies. More study is required to clarify the distribution of *Faxonella* in the Red River region near to northeastern Texas.

Orconectes

Along with giving background on this group, we now explain why we present six species of the genus *Orconectes*. The progress toward correct and comprehensive naming of the crawfish species has been slow and difficult. The taxonomy of *Orconectes*, however, has a particularly tortured past, providing as its legacy a good dose of uncertainty and confusion.

Shrimp crawfish

The first species that we present is the shrimp crawfish, *Orconectes lancifer*. Though a rare species, it has been known for over a century after being named by Hagen (1870). It has been collected from time to time in the Mississippi flood plain from near the Gulf to southmost Illinois, the Red River basin to Red River County in Texas and along the Gulf coastal plain to the Calcasieu River drainage north of Lake Charles, Louisiana. It is unique and little confusion exists about its species status.

Bluefingers

The next species grouping includes bluefingers: *palmeri, difficilis*, and *deanae*. The most common and widespread is *palmeri*, particularly the subspecies known as *palmeri longimanus*. Faxon (1898) based the type description on specimens from the Red River in Lamar County, Texas and southeastern Oklahoma. The specimens were submitted by S. E. Meek to where Faxon worked, the Museum of Comparative Zoology at Harvard Faxon considered the new animal as a variety of *Cambarus palmeri* (now in genus *Orconectes*) giving it the name *Cambarus palmeri longimanus*. Faxon, in 1884, described what is now known as *Orconectes palmeri palmeri* from a creek draining into Reelfoot Lake in Tennessee.

In 1933, E. P. Creaser and A. I. Ortenburger wrote a synopsis of the crustaceans of Oklahoma. They considered *palmeri longimanus* distinctive enough to treat with species rank (*Orconectes longimanus*). Crawfish scientists of following years did not adopt that view. Perhaps, this was because A. B. Williams (1954) gave an opinion that specimens he collected in central Arkansas were intergrades between *palmeri longimanus* and *palmeri palmeri*. Penn (1957) conducted a study of variation in the different subspecies (*palmeri palmeri*, *palmeri longimanus* and *palmeri creolanus*) that was used to establish geographical range for each. After examining specimens from the same area where the intergrades were found, Penn agreed with Williams that *palmeri longimanus* should be retained as a subspecies. Penn also established the range for *palmeri longimanus* as west of the Mississippi River in its tributaries, from the Arkansas River to the Gulf, and from there westward to the Guadalupe River in Texas. Though we intuitively consider that Creaser and Ortenburger's opinion may prove correct in the long run, we present all aforesaid varieties as one species, *Orconectes palmeri* (Faxon).

We collected several *palmeri longimanus* as mature males and females in the Trinity River drainage (Polk County). Reimer and Clark (1974) recorded one specimen from the Brazos basin in Robertson County. No mention was made as to the size or sex. Reddell (1965) reported specimens of *palmeri longimanus* from caves in Menard County, which could be considered part of the Colorado River basin. Penn and Hobbs (1958) recorded several Texas localities for *palmeri longimanus* including and between the Neches River and the Guadalupe River. Oddly, Hobbs (1990) later states that he is aware of no records for this crawfish in the Neches watershed. Fitzpatrick, in a 1987 paper that established subgenera for *Orconectes*, limited the range for *palmeri longimanus* and related forms in western Gulf drainages to no farther than the Neches basin. With only these localities and unspecified localities in Penn's work (1957) to draw upon for this crawfish in central and western Texas, we consider all reporting of *Orconectes palmeri longimanus* west of the Trinity to be of dubious records. It is possible that with the understanding of *Orconectes* at that time, and with many identifications being made with discolored museum specimens, early records were in some cases actually *palmeri longimanus* and in others, perhaps some other variety or species.

We present for the first time, two Texas forms that look very much like *Orconectes palmeri palmeri*. To distinguish them from the rest, we call these the Guadalupe River and Frio River forms. Bluish fingers are very obvious only in the smallest sizes. The Guadalupe form has a gonopod practically indistinguishable from *palmeri palmeri*. The Frio form has an appearance practically indistinguishable from the Guadalupe form except for the gonopod. Clearly, these two forms are geographically isolated from *palmeri palmeri* by a great distance. These crawfishes are apparently smaller than *palmeri longimanus*. The data in Penn's 1957 study dealing with *palmeri* subspecies shows that *Orconectes* specimens were examined from the Neches, Trinity, San Jacinto, Colorado and Guadalupe basins. From the locality map of that study and the later report by Penn and Hobbs (1958) on Texas crawfishes, it would appear that *palmeri longimanus* was recorded by them from Kimble and Gillespe County (Colorado basin: Llano River) and Kerr and Blanco Counties (Guadalupe basin: Guadalupe and Blanco Rivers). Reimer and Clark (1974), as mentioned above, reported it from the Brazos basin. The data of Penn (1957) indicates a single Form I form male from the Colorado basin and does not indicate a mature male from the Guadalupe. Penn and Hobbs (1958) do not give detailed specimen data.

In the same 1898 publication that Faxon named *palmeri longimanus*, he described *difficilis*. As he explained in that paper, *difficilis*, must be separated from *palmeri longimanus* on the basis of the gonopod because otherwise both species have the same general appearance. The gonopod's terminal elements are shorter in *difficilis* than *palmeri longimanus*. Jerry G. Walls (1972)described a new subspecies, *difficilis maletae,* from northwestern Louisiana.. Later, others elevated the subspecies to species rank (*maletae*). In the 1972 paper, Walls included a collection in Upshur County, Texas as a locality for his new variety. Based on our specimens collected in northeast Texas and comparisons to Walls' description, we consider his Texas record to be *Orconectes difficilis* (Faxon).

Reimer and Jester (1975) described *Orconectes deanae* from Conchas Lake, New Mexico. The upper reaches of South Canadian River drainage feed Conchas Lake. Subsequent work by Taylor, Jones and Bergey (2004) adds Canadian basin locations in New Mexico and Oklahoma. We report *deanae* from the same basin in Lake Marvin.

Virile crawfishes

The next group of *Orconectes* we will call virile crawfishes. We include two species in the Texas list. The first description of this type of crawfish was by Hermann Hagen in 1870. He described *virilis* from several North American locations including "Texas." Faxon (1885) repeated Hagen's record and in 1898 listed *virilis* from Lamar County. In their 1958 synopsis of Texas crawfishes, Penn and Hobbs considered these Texas records, including the one communicated in the original description, as erroneous. Faxon (1914) reported about *Cambarus virilis* and *Cambarus propinquus* (now known as *Orconectes virilis* and *Orconectes propinquus*) based on a collection taken in 1901 from Rio Conchos, at Jimenez, Province of Chihuahua, Mexico. The *virilis* specimens were all females. The *virilis* specimens so far mentioned were lost in the Chicago fire of 1871. In 1985, Ernesto Campos-Gonzalez and Salvador Contreras-Balderas reported on a collection of *virilis* from Rio San Pedro, Meoqui, Chihuahua. The Rio San Pedro is a tributary of the Rio Concho. They suggested that the presence of *virilis* may have been the result of its introduction by fishermen into the Rio Bravo (= Rio Grande).

Faxon described *Cambarus nais* (now known as *Orconectes nais*) in 1885 from Labette County, Kansas. This crawfish was subsequently reported from Oklahoma, Colorado, Arkansas, Missouri and Texas. Penn and Hobbs (1958) recorded two Form II males from Cooke County. A mention of the species in Texas was made by Raymond Bouchard and Henry Robison in a 1980 study of Arkansas crawfishes. In that paper, the authors provide a clarification that the Williams (1954) report of *nais* from Arkansas and Missouri were really *virilis*. William Pflieger, in a 1996 book on Missouri crawfishes, follows this clarification to not include *nais* in that state's fauna. Instead he portrays all such animals as *virilis*. Oklahoma records were contributed by Creaser and Ortenburger (1933) and Reimer (1969a).

Douglas Jester described a crawfish from Conchas Lake, New Mexico as *Orconectes causeyi* in 1967. Specimens had been sent to Horton Hobbs for identification and the response was that they represented a new species. Jester then described the crawfish as a new animal. Later it was discovered that animals examined by Hobbs were part of a mix of two species collected from the lake. Reimer and Jester would later (1975) describe one of these as *deanae*.

The form now understood as *causeyi* was later thought by Hobbs (1974, 1989) to be indistinguishable from *virilis*. In the meantime, Philip Unger (1978) reported on the crawfishes of Colorado. His work suggested that *causeyi* had greater similarity to *nais* than *virilis*. After examining virile crawfishes from Colorado and Missouri, and reviewing all depictions of these animal types in the literature, we agree with Hobbs that *causeyi* is probably indistinguishable from *virilis*.

Reimer (1969a) reports on the crawfishes of Oklahoma. In that work, Texas is listed as part of the *causeyi* distribution but no details are given. In his dissertation, Reimer (1969b) states that *causeyi* occurs "throughout the Pecos River" but again gives no details. We did not include the Pecos basin in the range map of *Orconectes* depicted earlier in this book.

The confusion concerning *Orconectes* of the mid-United States centers around the true status of *nais*. Original material, including the types, may have been lost. Researchers rely on early species descriptions that are only a few sentences in length. Gonopod depictions of name associations in subsequent literature are inconsistent. Perceptions differ as to which subgroup (*palmeri*-like bluefingers or *virilus*-like virile crawfishes) *nais* is best associated with. Both subgroups are quite similar in appearance. We infer that Reimer, Brouchard and Robison consider *nais* as a *palmeri*-like crawfish with an open areola, while Unger and Williams consider it a crawfish similar to *virilis*. J. F. Fitzpatrick's 1987 paper which proposed a series of subgenera for *Orconectes* did little to clarify the problem of *nais* but did associate the species closely with *virilis* and *causeyi*.

We found two crawfishes that fit structurally into our category of virile crawfishes. One is known to live in the upper Guadalupe River, a stream that courses through the Texas Hill Country. The other is found in the Brazos basin a few miles into the coastal plain from the Hill Country. We consider the crawfish we found in the Brazos basin to be *virilis*, and possibly a new subspecies. We consider the one in the Guadalupe basin to be *nais*, and possibly a new subspecies. By determining that crawfish to be *nais*, we are following Unger (1978) and others in assuming that *nais* is a virile crawfish and following Hobbs (1989) and others that *causeyi* is probably a synonym or a subspecies of *virilis*. Our designation of the Hill Country specimens as *nais* does little to clarify the confusion surrounding the name *Orconectes nais*. The specimens do conform to our concept of this species and the designation will serve our purpose until a taxonomic revision is done.

Other *Orconectes*

As reported by Penn and Hobbs (1958), Faxon (1898) recorded the presence of *Cambarus neglectus* Faxon 1885 (now *Orconectes neglectus*) from Arthur City (Lamar County), Texas. Penn and Hobbs (1958) included the record as part of the Texas fauna with some question, stating that the nearest presence known for this species is in northeastern Oklahoma (Arkansas River Drainage). Taylor, Jones and Bergey (2004) recorded *Orconectes neglectus* in the Red River Basin in Johnston County, Oklahoma, a location also far from northeastern Oklahoma. They suggest that the presence there may be the result of human introduction. We mention these reports here, but we do not include *Orconectes neglectus* in our list of Texas species.

It is possible or even likely that some presence of *Orconectes* in west and west-central Texas derives from natural, not introduced, stock. Our findings of multiple forms of virile crawfishes in west-central Texas, and even farther range extension of *Orconectes* into Mexico by Campos-Gonzalez and Contreras-Balderas, (1985) and Faxon (1914) indicate a sustained natural presence for the group. One could only guess that Hagen's and Faxon's earlier Texas listings for *virilis* could have been part of the collections that brought specimens for the type description of *Procambarus clarkii* by Girard in 1852. Girard's location for *clarkii* was "between San Antonio (Texas) and El Paso del Norte." It is probable that certain answers about the human role in the introduction of Texas crawfish species will never come to light. The example of *Orconectes*, however, does highlight a need to state a clear rationale when making assumptions about the human influence in species distribution.

Procambarus

Longpalms

The known longpalm species, *brazoriensis*, *hinei* and *incilis*, all occur in Texas. Ortmann (1905) described the first longpalm as *Cambarus hinei* (now *Procambarus hinei*) from southeastern Louisiana. Penn (1953b) redescribed the species, noting a number of Louisiana and Texas localities. Later, Penn (1962) and Albaugh (1975) described the other two longpalms from south-central Texas. These are well established species, anatomically distinct but close in general appearance.

Slenderclaws

The many slenderclaw species range over a large portion of the eastern and southern United States, through Texas and well into eastern Mexico. Three of the six species that we list for Texas occur only in that state. The range limit of a fourth species is unknown.

Clarification of species associations within the slenderclaw group took well over a century to reach the present understanding. Girard described the oldest member of the Texas group as *Cambarus acutus* (now *Procambarus acutus*) in 1852. Another species from the Carolinas, *Astacus Blandingii* Harlan, 1830 (now *Procambarus blandingii*) was considered a polytypic species (*b. blandingii* and *b. acutus*) after Faxon (1914) determined Girard's crawfish as a subspecies of *blandingii*. In 1941, Hobbs added another subspecies from San Luis Potosi, Mexico (*b. cuevachicae*) to the mix. For more than half of century and until it was recognized to be a truly distinct species, *Procambarus acutus* (Girard) was known as *Procambarus blandingii acutus* in Texas and elsewhere.

Hobbs (1967) established that *acutus* described longtime earlier by Girard (1852) was a true species. Hobbs thus separated Girard's crawfish from the Carolina species, again making *Procambarus blandingii* (Harlan) a distinct binomial species. The change also established two other trinomials. They became *Procambarus acutus acutus* (Girard) and *Procambarus acutus cuevachicae* (Hobbs).

Following Hobb's 1967 revision, four new species, *Procambarus texanus* Hobbs, 1971, *Procambarus nechesae* Hobbs, 1990, and *Procambarus zonangulus* Hobbs and Hobbs, 1990 and *Procambarus nueces* Hobbs and Hobbs, 1995 were described from Texas. We also notice a different form living in south Texas. In comments about the range of *Procambarus acutus acutus*, Hobbs (1989) mentions intergrading that occurs with *acutus acutus* and *acutus cuevachicae* in southwestern Texas and northern Mexico. The form that we see in south Texas does not appear to be an intergrade.

The description of *zonangulus* by Hobbs and Hobbs (1990) mentions a close relationship of that species with *texanus*. Slenderclaw specimens that we caught from the lower Colorado and Brazos basins had gonopod features characteristic of both species. We are hesitant to say that they represent an intergrading of the two species, because of the limited number of specimens examined, and because the gonopods more closely conform to *texanus*.

Red crawfish

The red crawfish, *Procambarus clarkii* (Girard) 1852, is the first of its subgroup to receive a name. It has a broad geographic range. The original description of *clarkii* was based on specimens collected between San Antonio and El Paso.

Southern streamers

The two known southern streamers of Texas, *natchitochae* and *dupratzi* were described by George H. Penn (1953a). The original description of *dupratzi* was based on a specimen collected in Shelby County, Texas.

Grays

We list three gray crawfishes, *simulans*, *curdi* and *regiomontanus,* for Texas. Faxon described the oldest Texas gray, *simulans*, from Dallas, Texas specimens in 1884. The only gray crawfish presented by Penn and Hobbs in their 1958 review of Texas crawfishes was *Procambarus simulans simulans* (Faxon). They portrayed *simulans* as one of the state's most common crawfishes. They reported on specimens examined from the Rio Grande basin and eastward and northward from the Colorado River Basin. Included in the review was mention that *simulans simulans* specimens collected from Longhorn Cavern in Burnet County were examined and other crawfishes had been seen in other Texas caves. The trinomial was used for *simulans* by Hobbs and Penn because Alejandro Villalobos had, in 1954, established *Procambarus simulans regiomontanus* as a new subspecies for a gray crawfish inhabiting the province of Nuevo Leon, Mexico. Hobbs and Robison (1988) proposed that the Mexican crawfish should be given species status as *Procambarus regiomontanus* Villalobos. If this proposal is accepted, it would make *simulans* binomial again.

An unpublished dissertation by Rollin Reimer (1969b), explained the status of *simulans*-related crawfishes in Texas and elsewhere. Reimer (1975) later described a gray crawfish, *curdi*, from Texas, Oklahoma and Arkansas. Specimens from Brazos County, Texas served as the description types.

Reimer's dissertation (1969b) also described a crawfish species with two forms from Texas that was similar in appearance to *simulans* but distinguishable from it in gonopod structure. The species is described as having a central projection that is longer than the caudal process (not so in *simulans*). This, or a very similar crawfish from New Mexico, had been recognized and described earlier by T. Cockerell and W. Porter (1900). Reimer did not accept the species named therein because he found , in the depository's bottle, a *simulans* specimen representing the type.

Hobbs (1989) lists the species name *Cambarus gallinus* Cockerell and Porter (1900) as a synonym of *Procambarus simulans* (Faxon). Reimer (1969b) acknowledged that Cockerell and Porter did recognize a new species. Based on our collections and the reports mentioned above, we understand that the distribution of such a crawfish with elongated central projection overlaps that of typical *simulans* in Texas. It is more likely to be encountered in western and northwestern parts of the state, including the panhandle. Specimens of this kind that we collected in the Pecos basin in Carlsbad, New Mexico have a bolder pattern of dark markings than specimens from west-central Texas. Future work may determine that this crawfish form is only a subspecies of *simulans*. Presently, it is included with *simulans* in the published literature. We enter it in our list as a form of *simulans.*

The third gray lives in far south Texas. It has been known for sometime as a species close to, but distinct from, *Procambarus regiomontanus* Villalobos. The history of *regiomontanus* includes its introduction to the nomenclature by Villalobos (1954) as a subspecies of *simulans* (see above) from a rather localized area in Nuevo Leon. Subsequently, the geographic range of *regiomontanus* was better established within Nuevo Leon and Tamaulipas by Rodriguez-Almaraz, Coronado-Magdaleno, and Campos-Gonzales (1993) and Rodriguez-Almaraz and Campos-Gonzales (1994). Reimer (1969b) reported on specimens that he examined from southern Texas. He identified these (part tentative) as *regiomontanus* and listed Dimmit, Hidalgo and Zavala Counties as the collection localities. Considering our collections and the work of others, we list the crawfish that lives in the lower Rio Grande (Rio Bravo) of Texas as *regiomontanus*. The depictions of Villalobos and others that followed of the *regiomontanus* gonopod show, in our opinion, a gonopod that is not oriented as a true lateral view. Differences in those records and our figure are a result of perspective. The gonopod of the south Texas specimens may be rotated to appear exactly as shown in the earlier depictions of a *regiomontanus* gonopod.

Mexican scientists report an undescribed species from the Mexico side of the lower Rio Bravo (Rio Grande) basin. It is said by them to be very similar to *regiomontanus*. Perhaps the species we call *regiomontanus* and catch in the Texas side of this basin is the same animal. We cannot determine this without further study of specimens. Our specimens, however, do conform to the characteristics of *regiomontanus*.

Creekbank crawfish

A rather unique species, *Procambarus kensleyi*, was described as part of a survey of the Neches River basin (Hobbs 1990). Hobbs indicates that this species is probably the one originally described in Reimer's dissertation.

Hairclaws

Procambarus nigrocinctus was described by Hobbs (1990). His material was collected in a survey of Neches River crawfishes. Reimer had collected this species and described it (without giving a binomial name) in his dissertation (1969b). In natural classification, this and the nearby hairclaw species, *Procambarus tulanei* Penn (1953c), would both be included with what we call gray crawfishes. We artificially separate them as hairclaws using the beard as the distinctive character.

Prairie crawfishes

The prairie crawfishes have scant record in Texas. The first accepted record was that of *Procambarus gracilis* (Bundy) from Cooke County by Penn and Hobbs (1958). It consisted of one female. We infer that this specimen was examined again by Reimer (1969b) and that he included it in the material used for the description of a new species (*Procambarus* species B). He said that the range of *P.* species B was the northwest of Cooke County, Texas into northern Arkansas. That species description probably defined a crawfish that was later published as *Procambarus liberorum* by Fitzpatrick (1978). The true identity of the Texas specimen is conjectural, considering the description of the following Texas species. In 1991, Hobbs described a new species, *Procambarus steigmani*, from a long-grass prairie in Collin County. This species was explained as very similar in appearance to a species, *Procambarus regalis* Hobbs and Robison, 1988, previously described from southwest Arkansas. We collected both types in Texas; only one specimen was a Form I male. This male *regalis* was collected in Fannin County. It was colorful like *steigmani* but, unlike a feature said by Hobbs to be diagnostic for *steigmani*, the arrangement of its gonopods was symmetrical.

Navasota crawfish

Procambarus caeruleus Fitzpatrick and Wicksten, 1998 was initially described using a few juvenile specimens in a dissertation by Reimer (1969b). He recognized that the gonopod and other characters were sufficiently unique to merit, not only a status of new species, but also a special grouping within the *Procambarus*. Later, Fitzpatrick and Wicksten (1998), also using a limited number of specimens, published a description of the crawfish from the same county (Brazos) and river basin (Navasota). They did not mention Reimer's study. We possess a good series of Form I males and agree with Reimer that this crawfish is especially unique among the crawfishes.

Future study will better establish the ranges of Texas crawfishes and probably determine several new species. Scientific collections have not been made in many counties. Collectors have never visited many habitats that are suitable for crawfish life, such as wet areas far from permanent waters, isolated seeps and springs and banks and rocky places of larger rivers.

Checklist of Texas Crawfishes

Pygmies
- *Cambarellus ninae* Hobbs, 1950
- *Cambarellus puer* Hobbs, 1945
- *Cambarellus shufeldtii* (Faxon) 1884
- *Cambarellus texanus* Albaugh and Black, 1973

Fatclaws
- *Cambarus diogenes* Girard, 1852 - 2 varieties
- *Fallicambarus devastator* Hobbs and Whiteman,1987
- *Fallicambarus foidens* (Cottle) 1863
- *Fallicambarus macneesei* (Black) 1967

Grippers
- *Faxonella beyeri* (Penn) 1950
- *Faxonella blairi* Hayes and Reimer, 1977
- *Faxonella clypeata* (Hay) 1899

Shrimp Crawfish
- *Orconectes lancifer* (Hagen) 1870

Bluefingers
- *Orconectes deanae* Reimer and Jester, 1975
- *Orconectes difficilis* (Faxon) 1898
- *Orconectes palmeri* (Faxon) 1884 - 3 forms

Virile crawfishes
- *Orconectes nais* (Faxon) 1885
- *Orconectes virilis* (Hagen) 1870

Longpalms
- *Procambarus brazoriensis* Albaugh, 1975
- *Procambarus hinei* (Ortmann) 1905
- *Procambarus incilis* Penn, 1962

Slenderclaws
- *Procambarus acutus* (Girard) 1852
- *Procambarus nechesae* Hobbs, 1990
- *Procambarus nueces* Hobbs and Hobbs, 1995
- *Procambarus texanus* Hobbs, 1971
- *Procambarus zonangulus* Hobbs and Hobbs, 1990
- South Texas slenderclaw, undescribed

Red Crawfish
- *Procambarus clarkii* (Girard) 1852

Streamers
- *Procambarus dupratzi* Penn, 1953
- *Procambarus natchitochae* Penn, 1953

Grays
- *Procambarus curdi* Reimer, 1975
- *Procambarus regiomontanus* Villalobos, 1954
- *Procambarus simulans* (Faxon) 1884 - 2 forms

Creekbank Crawfish
- *Procambarus kensleyi* Hobbs, 1990

Hairclaws
- *Procambarus nigrocinctus* Hobbs, 1990

Prairies
- *Procambarus regalis* Hobbs and Robison, 1988
- *Procambarus steigmani* Hobbs, 1991

Navasota Crawfish
- *Procambarus caeruleus* Fitzpatrick and Wicksten, 1998

State total: 37 species, one of which is unnamed and three species of which consist of two or three varieties or forms.

Literature Cited

Albaugh, Douglas W. 1975. A new crawfish of the genus *Procambarus*, subgenus *Capillicambarus*, from Texas, with notes on the distribution of the subgenus. *Tulane Studies in Zoology and Botany*, 19(1,2):1-7.

Albaugh, Douglas W., and Joe B. Black. 1973. A new crawfish of the genus *Cambarellus* from Texas, with new distributional records for the genus (Decapoda, Astacidae). *Southwestern Naturalist*, 18(2):177-185.

Black, Joe B. 1967. A new crawfish of the genus *Cambarus* from southwest Louisiana (Decapoda: Astacidae). *Proceedings of the Biological Society of Washington*, 80:173-178.

Bouchard, Raymond W., and Henry W. Robison. 1980. An inventory of the decapod crustaceans (crayfishes and shrimps) of Arkansas with a discussion of their habitats. *Arkansas Academy of Science Proceedings*, 34:22-30.

Campos-Gonzales, Ernesto, and Salvador Contreras-Balderas. 1985. First record of *Orconectes virilis* (Hagen) (Decapoda, Cambaridae) from Mexico. *Crustaceana*, 49(2):218-219.

Cockerell, T., and W. Porter. 1900. A new crayfish from New Mexico. *Proceedings of the Academy of Natural Sciences of Philadelphia*, 52:434-435.

Cottle, T. J. 1863. On the two species of *Astacus* found in upper Canada. *Canadian Journal of Industry, Science, and Arts, new series*, 45:216-219.

Creaser, E. P., and A. I. Ortenburger. 1933. The decapod crustaceans of Oklahoma. *Publications of the University of Oklahoma Biological Survey*, 5(2):14-47.

Faxon, Walter. 1884. Descriptions of new species of *Cambarus*; to which is added a synonymical list of the known species of *Cambarus* and *Astacus*. *Proceedings of the American Academy of Arts and Sciences*, 20:107-158.

Faxon, Walter. 1885. A revision of the Astacidae, Part I: The genera *Cambarus* and *Astacus*. *Memoirs of the Museum of Comparative Zoology at Harvard College*, 10(4): vi + 186 pages, 10 plates.

Faxon, Walter. 1898. Observations on the Astacidae in the United States National Museum and in the Museum of Comparative Zoology, with descriptions of new species. *Proceedings of the United States National Museum*, 20(1136):643-694, plates 62-70.

Faxon, Walter. 1914. Notes on the crayfishes in the United States National Museum and the Museum of Comparative Zoology with descriptions of new species and subspecies to which is appended a catalogue of the known species and subspecies. *Memoirs of the Museum of Comparative Zoology at Harvard College*, 40(8):351-427, 13 plates.

Fitzpatrick, J. F. 1978. A new crawfish of the subgenus *Girardiella*, genus *Procambarus* from northwest Arkansas (Decapoda: Cambaridae). *Proceedings of the Biological Society of Washington*, 91(2):533-538.

Fitzpatrick, J. F. 1987. The subgenera of the crawfish genus *Orconectes* (Decapoda: Cambaridae). *Proceedings of the Biological Society of Washington*, 100(1):44-74.

Fitzpatrick, J. F., and M. K. Wicksten. 1998. A new crawfish of the genus *Procambarus* (Crustacea: Decapoda: Cambaridae) from central Texas. *Proceedings of the Biological Society of Washington*, 111(1):146-152.

Girard, Charles. 1852. A revision of the North American Astaci, with observations on their habits and geographical distribution. *Proceedings Academy of Natural Sciences Philadelphia*, 6:87-91.

Hagen, Hermann A. 1870. Monograph of the North American Astacidae. *Illustrated Catalogue of the Museum of Comparative Zoology at Harvard College*, No. 3, viii + 1-111,11 plates.

Harlan, R. 1830. Description of a new species of the genus *Astacus*. *Transactions of the American Philosophical Society*, 3(15):464-465.

Hay, William P. 1899. Description of two new species of crayfish. *Proceedings of the United States National Museum*, 22(1187):121-123.

Hayes II, William A., and Rollin D. Reimer. 1977. *Faxonella blairi*, a new crawfish from the Red River Drainage of Oklahoma and Arkansas. *Proceedings of the Biological Society of Washington*, 90(1);1-5.

Hobbs, Horton H. Jr. 1941. A new crayfish from San Luis Potosi, Mexico(Decapoda: Astacidae). *Zoologica*, 26(1):1-4.

Hobbs, Horton H. Jr. 1945. Two new species of crayfishes of the genus *Cambarellus* from the gulf coastal states, with a key to the species of the genus. *American Midland Naturalist*, 34(2):334-357.

Hobbs, Horton H. Jr. 1948. A new crayfish of the genus *Cambarus* from Texas, with notes on the distribution of *Cambarus foidens* (Cottle). *Proceedings of the United States National Museum*, 98(3230):223-231.

Hobbs, Horton H. Jr. 1950. A new crayfish of the genus *Cambarellus* from Texas (Decapoda, Astacidae). *Proceedings of the Biological Society of Washington*, 63(18):89-96.

Hobbs, Horton H. Jr. 1967. The current status of the crayfishes listed by Girard (1852) in his "A revision of the North American Astaci...." *Crustaceana*, 12(2):124-132.

Hobbs, Horton H. Jr. 1969. On the distribution and phylogeny of the crayfish genus *Cambarus. In* Perry C. Holt, Richard L. Hoffman, and Willard Hart Jr., The Distributional History of the Biota of the Southern Appalachians, Part I: Invertebrates. *Virginia Polytechnic Institute, Research Division Monograph*, 1:93-178, figures 1-20.

Hobbs, Horton H. Jr. 1971. New crayfishes of the genus *Procambarus* from Alabama and Texas (Decapoda: Astacidae). *Proceedings of the Biological Society of Washington*, 84(11):81-94.

Hobbs, Horton H. Jr. 1974. A checklist of the North and Middle American crayfishes (Decapoda: Astacidae and Cambaridae). *Smithsonian Contributions in Zoology*, number 166, 161 pages.

Hobbs, Horton H. Jr. 1981. The crayfishes of Georgia. *Smithsonian Contributions in Zoology*, number 318, 549 pages.

Hobbs, Horton H. Jr. 1989. An illustrated checklist of the American crayfishes (Decapoda: Astacidae, Cambaridae, and Parastacidae). *Smithsonian Contributions to Zoology*, number 480, 236 pages.

Hobbs, Horton H. Jr. 1990. On the crayfishes (Decapoda: Cambaridae) of the Neches River basin of eastern Texas with the descriptions of three new species. *Proceedings of the Biological Society of Washington*, 103(3):573-597.

Hobbs, Horton H. Jr. 1991. *Procambarus (Girardiella) steigmani* (Decapoda: Cambaridae), a new crayfish from a long grass prairie in northeastern Texas. *Proceedings of the Biological Society of Washington*, 104(2):309-316.

Hobbs, Horton H. Jr., and Horton H. Hobbs III. 1990. A new crayfish from southeastern Texas. *Proceedings of the Biological Society of Washington*, 103(3):608-613.

Hobbs, Horton H. Jr., and H. H. Hobbs III. 1995. *Procambarus (Ortmannicus) nueces* (Decapods: Cambaridae), a new crayfish from the Nueces River Basin, Texas. *Proceedings of the Biological Society of Washington*, 108(1):54-60.

Hobbs, Horton H. Jr., and H. W. Robison. 1988. The crayfish subgenus *Girardiella* (Decapoda: Cambaridae) in Arkansas, with the descriptions of two new species and a key to members of the *gracilia* Group of the genus *Procambarus. Proceedings of the Biological Society of Washington,* 101(2):391-413.

Hobbs, Horton H. Jr., and H. W. Robison. 1989. On the crayfish genus *Fallicambarus* (Decapoda: Cambaridae) in Arkansas, with notes on the *foidens* complex and descriptions of two new species. *Proceedings of the Biological Society of Washington*, 102:651-697.

Hobbs, Horton H. Jr., and Mike Whiteman. 1987. A new, economically important crayfish (Decapoda: Cambaridae) From the Neches River basin, Texas, with a key to the subgenus *Fallicambarus. Proceedings of the Biological Society of Washington*, 100(2):403-411.

Hobbs, Horton H. Jr., and Mike Whiteman. 1991. Notes on the burrows, behavior, and color of the crawfish *Fallicambarus (F.) devastator* (Decapoda: Cambaridae). *The Southwestern Naturalist*, 36(1):127-135.

Jester, Douglas B. 1967. A new crawfish of the genus *Orconectes* from New Mexico (Decapoda, Astacidae). *American Midland Naturalist*, 77(2):518-524.

Ortmann, A. E. 1905. A new species of *Cambarus* from Louisiana. *Ohio Naturalist*, 6(2):401-403.

Penn, George H. 1950. A new crawfish of the genus *Orconectes* from Louisiana (Decapoda: Astacidae). *Journal of the Washington Academy of Sciences*, 40(5):166-169.

Penn, George H. 1953a. Two new crawfishes of the genus *Procambarus* from Texas, Louisiana, and Arkansas (Decapoda, Astacidae). *American Museum Novitates*, 1636:1-10.

Penn, George H. 1953b. A redescription of the crawfish *Procambarus hinei* (Ortmann). *Tulane Studies in Zoology*, 1(5):63-68.

Penn, George H. 1953c. A new crawfish of the Genus *Procambarus* from Louisiana and Arkansas (Decapoda: Astacidae). *Journal of the Washington Academy of Sciences*, 43(5):163-166.

Penn, George H. 1957. Variation and subspecies of the crawfish *Orconectes palmeri. Tulane Studies in Zoology*, 7(1):3-20.

Penn, George H., and Horton H. Hobbs Jr. 1958. A contribution toward a knowledge of the crayfishes of Texas (Decapoda, Astacidae). *Texas Journal of Science*, 10(4):452-483.

Penn, George H. 1962. A new crawfish of the *Hinei* Section of the genus *Procambarus* (Decapoda: Astacidae). *Crustaceana*, 3(3):222-226.

Pflieger, William L. 1996. *The crawfishes of Missouri.* Missouri Department of Conservation, Jefferson City, 152 pages.

Reddell, James R. 1965. A checklist of the cave fauna of Texas I. The Invertebrata (exclusive of Insecta). *The Texas Journal of Science*, 17(2):143-187.

Reimer, Rollin D. 1969a. A report on the crawfishes (Decapoda, Astacidae) of Oklahoma. *Proceedings of the Oklahoma Academy of Sciences*, 48:49-65.

Reimer, Rollin D. 1969b. A taxonomic study of the *Gracilis* section of the genus *Procambarus.* Dissertation, Tulane University, 190 pages.

Reimer, Rollin D. 1972. Notes on the genus *Faxonella* in Texas. *The Texas Journal of Science*, 24(2):263-264.

Reimer, Rollin D. 1975. *Procambarus* (*Girardiella*) *curdi*, a new crawfish from Arkansas, Oklahoma, and Texas (Decapoda, Astacidae). *Tulane Studies in Zoology and Botany*, 19 (1,2):22-25.

Reimer, Rollin D., and William Clark. 1974. Decapod crustaceans of the Navasota River system in central Texas. *Southwestern Naturalist*, 19(2):167-178.

Reimer, Rollin D., and Douglas B. Jester. 1975. A new crawfish of the genus *Orconectes* from Conchas Lake, New Mexico. *Tulane Studies in Zoology and Botany*, 19:17-21.

Reimer, Rollin D., Kenneth W. Johnson, and Rocco A. Marcello. 1975. The crawfish, *Fallicambarus macneesei* (Black) in Texas. *Texas Journal of Science*, 26(1&2):291.

Rodriguez-Almaraz, Gabino A., and Ernesto Campos-Gonzales. 1994 . Distribution and status of the crayfishes (Cambaridae) of Neuvo Leon, Mexico. *Journal of Crustacean Biology*, 14(4):729-735.

Rodriguez-Almaraz, Gabino A., Mario Coronado-Magdaleno, and Ernesto Campos-Gonzales. 1993. Distribucion y notas ecologicas de los acociles (Cambaridae: *Procambarus*) del estado de Tamaulipas, Mexico. *The Southwestern Naturalist*, 38(4):390-393.

Rogers, Robert, and Jay V. Huner. 1985. Comparison of burrows and burrowing behavior of five species of cambarid crawfish (Crustacea, Decapoda) from the Southern University Campus, Baton Rouge, Louisiana. *Proceedings of the Louisiana Academy of Science*, 8:23-29.

Taylor, C. A., S. N. Jones and E. A. Bergey. 2004. Crayfishes of Oklahoma revisited: New state records and checklist of species. *The Southwestern Naturalist*, 49(2):250-255.

Unger, Philip A. 1978. The crawfishes (Crustacea: Cambaridae) of Colorado. In *Natural History Inventory of Colorado*, number 3, University of Colorado Museum, Boulder, 19 pages.

Villalobos, Alejandro. 1954. Estudios de los cambarinos mexicanos, XI: Una nueva subspecie de *Procambarus simulans* del estado de Nuevo Leon. *Anales del Instituto de Biologia, Universidad Nacional Autonoma de Mexico*, 25(1,2):289-298, 2 plates.

Walls, Jerry G. 1972. Three new crawfishes related to *Orconectes difficilis* (Faxon) (Decapoda: Astacidae). *Proceedings of the Biological Society of Washington*, 84(53):449-458.

Wetzel, James E. 2002. Form alternation of adult female crayfishes of the genus *Orconectes* (Decapoda: Cambaridae). *American Midland Naturalist*, 147:326-337.

Williams, A. B. 1954. Speciation and distribution of the crayfishes of the Ozark Plateaus and Ouachita Provinces. *University of Kansas Science Bulletin*, 36(2):803-918.

English and Metric Units

Equivalents

1 inch = 2.54 centimeters
1 foot = 0.305 meter = 30.5 centimeters
1 mile = 1.609 kilometer
1 pound = 0.454 kilogram
1 ounce (U.S fluid) = 29.57milliliters

Conversions	
inches to centimeters	X 2.54
feet to meters	X 0.305
miles to kilometers	X 1.609
pounds to kilograms	X 0.454
ounces to milliliters	X 29.57

Subgenus Designations of Texas Crawfishes

For the sake of simplicity, subgenera were not included in the species presentations of this book. Because they are of some importance to biologists and helpful in computer searches, we associate below the subgenus names (in parenthesis) with the names given for Texas crawfishes in the book text.

Cambarellus (Pandicambarus)	The Texas pygmies except *Cambarellus shufeldtii*
Cambarellus ((Dirigicambarus)	*Cambarellus shufeldtii*
Cambarus (Lacunicambarus)	*Cambarus diogenes*
Fallicambarus (Fallicambarus)	*Fallicambarus devastator, F. macneesei*
Fallicambarus (Creaserinus)	*Fallicambarus foidens*
Orconectes (Buannulifictus)	*Orconectes palmeri*
Orconectes (Hespericambarus)	The other bluefingers
Orconectes (Gremicambarus)	The virile crawfishes
Orconectes (Tragulicambarus)	The shrimp crawfish

Procambarus (Capillicambarus)	The longpalms
Procmbarus (Ortmannicus)	The slenderclaws
Procambarus (Scapulicambarus)	The red crawfish
Procambarus (Pennides)	The southern streamers
Procambarus (Girardiella)	The grays
Procambarus (Girardiella)	The creekbank crawfish
Procambarus (Girardiella)	The Texas hairclaw
Procambarus (Girardiella)	The prairie crawfishes
Procambarus (Girardiella)	The Navasota crawfish
No subgenus designations	The grippers, *Faxonella*

Tips for crawfish conservation

When collecting specimens from a place, take only one or two specimens of each species. The removal of only one or two crawfish of a species assures survival of the crawfish population at that site. The practice of catch and release is another way to avoid over-crawfishing a locality.

Be sure that specimens are returned to the exact place of capture when released. By doing so, the natural distribution of a species is not disrupted.

Similarly, avoid dumping unused bait crawfish into places where they may continue to live and reproduce. Instead, freeze the crawfish for later use as frozen bait or chum.

Use common sense to avoid inappropriate releases. Unintentional transfers and accidental escapes may lead to the establishment of a crawfish beyond its natural range. The new presence of a species in a place is often detrimental to native crawfish species.

Avoid involvement in the sale or trade of a rare species. Commercialization puts the species at risk.

Follow any laws and regulations meant to protect the crawfish resource.

COUNTY LOCATOR

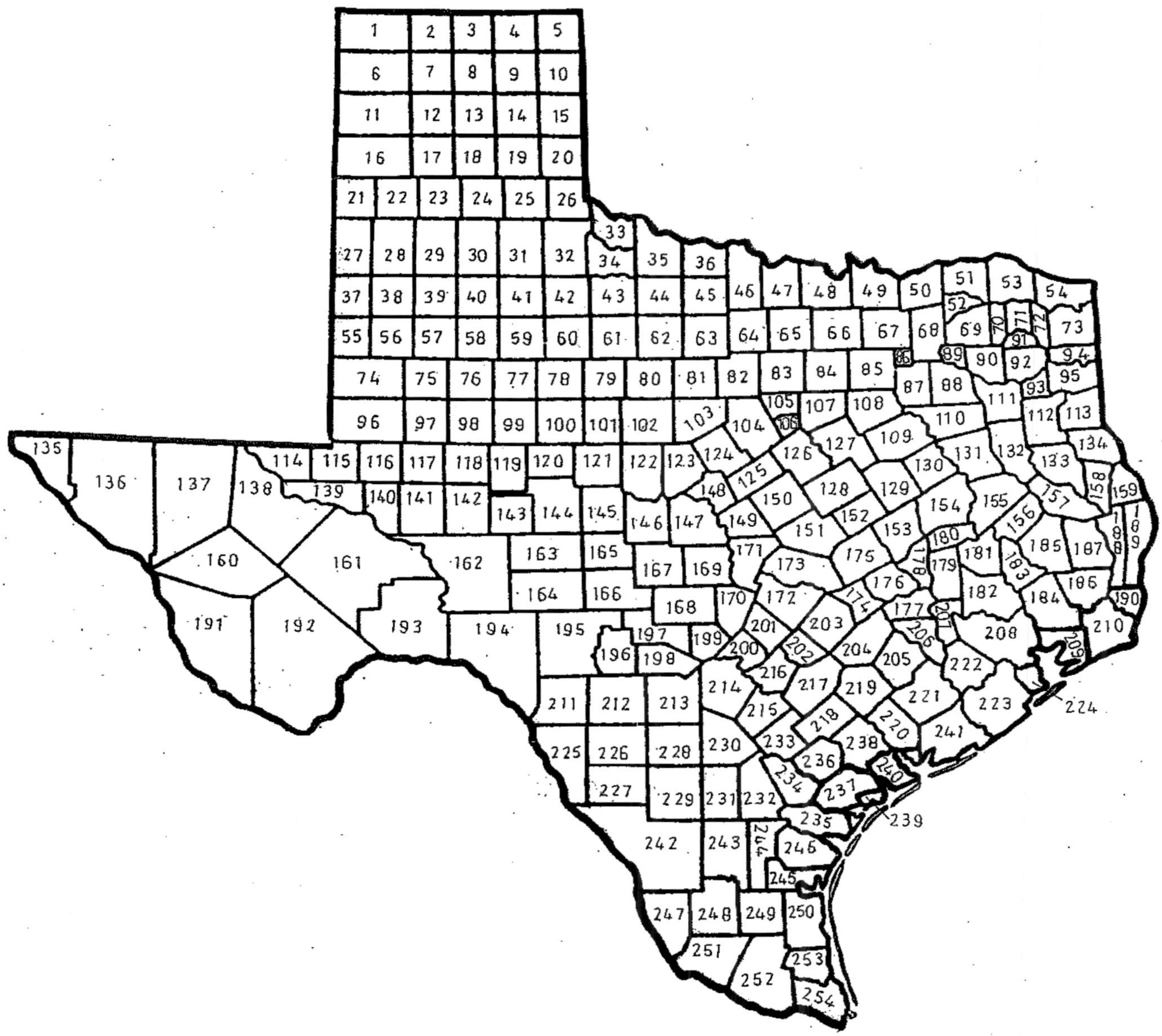

Texas counties.

Index

Sterling Ken Johnson (right) is a retired professor and aquatic animal disease specialist from Texas A&M University. Nathan Kurt Johnson (left) is owner/operator of Wet Environmental, a water resources business servicing central Texas. Ken resides with his wife, Mary Lou, in College Station and Kurt with his wife, Becky, in the San Marcos area. Both enjoy being outdoors, close to nature and, of course, talkin' crawdad.